Chaos Theory.
A Quick Immersion

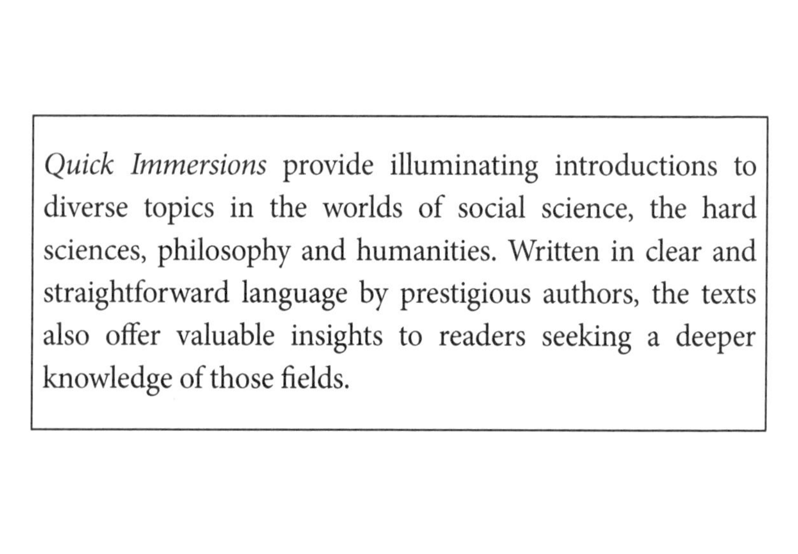
Quick Immersions provide illuminating introductions to diverse topics in the worlds of social science, the hard sciences, philosophy and humanities. Written in clear and straightforward language by prestigious authors, the texts also offer valuable insights to readers seeking a deeper knowledge of those fields.

Robert C. Bishop

CHAOS THEORY

A Quick Immersion

Tibidabo Publishing
New York

Copyright © Robert C. Bishop

Published by Tibidabo Publishing, Inc. New York.

All rights reserved. No part of this publication may be reproduced, stored in a retrieval system, or transmitted, in any form or by any means, electronic, mechanical, photocopying, recording, scanning or otherwise, without the prior permission in writing with the Publisher or as permitted by law, or under terms agreed upon with the appropriate reprographics rights organization.

Copyediting by Lori Gerson

Cover art by Raimon Guirado

All illustrations are by © Robert C. Bishop, except Figure 1 which is public domain but is attributed to NOAA.

First published: July, 2023

Visit our Series on our Web:

www.quickimmersions.com

ISBN: 978-1-949845-35-8

1 2 3 4 5 6 7 8 9 10

Library of Congress Control Number: 2023939935

Printed in the United States of America.

Advanced Praise

"Books on chaos theory tend to fall one of two ways. The first are those that cannot resist the hype. The exotic sound of "chaos," like "artificial intelligence," invites authors to lean into its mysteries. The second group of books is overly technical and beyond the reach of most readers. Physicists and mathematicians especially find it hard not to include all sorts of arcane detail. This book avoids both pitfalls. Robert Bishop deftly takes the reader through each of the key topics raised by chaos, including its historical roots, and shows why so many disciplines take note. Mathematicians provide fractal geometry as a tool. Computer scientists reveal how chaos puts limits on numerical simulations. Physical scientists have discovered that chaos is ubiquitous in nature. And philosophers explore its implications for a range of longstanding questions. With one foot in physics and another in the philosophy of science, Bishop illustrates each using minimal mathematics with important technical terms explained throughout. There is no better introduction to the topic available and no scholar better suited to the task."

Jeffrey Koperski, Professor of Philosophy, Saginaw Valley State University

"Sensitivity to initial conditions and path dependence, nonlinearities, strange attractors, and fractals...Bishop explains – in enviably clear language – the surprising properties and emergence of chaotic dynamics in mathematical models and real-world systems, and the differences between the two. The last chapter, on the limits chaos places on knowledge and predictability, provides insights on how individuals as well as scientists can work within those limits while at the same time embracing the remarkable wisdom they offer. Highly recommended."

Alicia Juarrero, author of Context Changes Everything: How Constraints Create Coherence (MIT Press)

"It won't come as a surprise to anyone who has been following Professor Bishop's work over the years on this topic that this book represents the best primer on chaos theory and nonlinear dynamics in existence today."

Michael Silberstein, Professor of Philosophy, Elizabethtown College

"Chaos Theory: A Quick Immersion is a clear and engaging introduction to chaos theory. Assuming no prior knowledge, and using helpful analogies and examples from everyday life, it familiarizes readers with key concepts and findings of this fascinating field. The discussion is readily accessible to those new to the topic, yet without skirting over important nuances. In addition to surveying conceptual foundations, Chaos Theory nicely illustrates how ideas and tools from the mathematical study of chaos have been applied in science – in weather forecasting, ecology, physiology, physics, and more. Along the way, it calls attention to oft-overlooked challenges involved in relating mathematical and computer models to the physical world and emphasizes the importance of recognizing and navigating limits when seeking knowledge. This delightful little book will be useful to a wide range of readers interested in understanding what chaos theory is and how its insights can make a difference in science."

Wendy Parker, Professor of Philosophy, Virginia Polytechnic University

Contents

Acknowledgments

I have benefited greatly from conversations over the years with Harald Atmanspacher, Fred Kronz, Tim Palmer, and Lenny Smith on the topics of chaos and nonlinear dynamics. Although formally distinct from my Stanford Encyclopedia of Philosophy article on chaos, it is hard to pull apart the influence of past writings on such a fascinating topic when revisiting it with fresh eyes.

Introduction

Chaos is a word with many meanings in our everyday world. We often feel we experience chaos when our day is jumbled and disorganized, when things keep happening that throw us off balance. We may feel that achieving our goals for the day were blocked by the unforeseen events that threw off our plans. Then there is the Marvel universe enemy organization, Chaos, which funds armies and countries to lead the world down a pathway to a particular future. Here, disorganization is deliberately sown to achieve a plan. Or we often use the word chaos as synonymous with randomness, as lawlessness, a complete lack of order or pattern.

By contrast, there is chaos as mathematicians and scientists talk about: small changes producing surprisingly large effects that appear unpredictable yet exhibit an exquisite kind of order. The **butterfly effect** illustrates this extreme sensitivity to tiny changes. The term originates from the idea that the flapping of a butterfly's wings in Brazil could cause a tornado in Texas three weeks later.

You might be thinking, "So what? We've always known that small changes can have big effects. What's special about this mathematical sense of chaos?" For instance, on July 7th, 2005, because of a delay of just a few seconds, a friend and her family missed boarding one of the London Underground trains attacked by a terrorist bombing. Like the organization Chaos, such terrorist attacks are intended to sow chaos and fear by violently disrupting ordinary life.

Understanding what is distinctive about the phenomenon and properties of chaos as mathematicians and scientists characterize and study it is what this book is about. The mathematical sense of chaos—often called **chaotic dynamics**—looks random at first glance but represents a surprising kind of order. This book will help you understand this surprising kind of order and how sensitivity to the smallest of changes has revolutionized how scientists think about the behavior of our world.

Randomness

Before we begin that journey, it is important to clear up one confusion about randomness or random behavior. In everyday talk, our tendency is to use the word random to mean a lawless or completely unordered behavior. Scientists never use the term random to mean this for an important reason: There are no examples of lawless disorder in any of the physical phenomena we study.

Confusion arises because systems behaving randomly appear to lack any order when we're watching their behavior. Scientists call this **apparent randomness** when a system looks random to us but has an underlying deterministic order to it. Think of a roulette wheel. The outcome of each spin with the ball landing in a particular numbered slot looks like there is no order. Yet suppose we were able to know the speed of the wheel's spinning, the initial velocity of the ball as it enters the wheel, the friction slowing the wheel down, the friction the ball experiences as it rolls around the wheel and eventually bounces into a slot, among several other factors. Given these factors, the final slot the ball settles in is fully determined. We might not be able to calculate this due to the many factors involved and the limits on our knowledge, but there is an underlying order to the system determining where the ball will land. It appears random to us because we cannot track all the factors involved. Nevertheless, the ball's behavior is fully determined in an ordered way.

There is a second form of randomness scientists study known as **irreducible randomness**. When the full set of physical conditions determine the probability for outcomes, but not the specific outcome in a system at a particular time, it is irreducibly random. Nonetheless, the irreducible randomness of these outcomes still conforms to fixed probabilities. These probabilities are constrained by statistical laws rather than deterministic laws. This means irreducible randomness is a different form of order than the deterministic order we experience with mechanical systems such as engines and computers. It definitely is not lawless chaos.

An example of irreducible randomness is radioactive decay. All the relevant factors in a sample of a radioactive element, such as uranium, will not determine when any specific nucleus in the sample will undergo a decay event. Nevertheless, the sample will behave as described by a statistical law constraining how many nuclei will decay on average during a given time interval. Scientists make use of such irreducible randomness all the time in medical treatments for cancer and in nuclear power plants.

Over the course of this book, you will see how these two forms of scientific randomness intersect with chaotic dynamics. Moreover, you will see that perhaps the most important lesson of chaos is that of limits, the limits of what can be forecast and the wisdom of working within these limits.

With this brief introduction, our journey begins.

Chapter 1

Sensitive Dependence

I began approaching chaotic dynamics with the idea of very small changes now producing very large effects in the future. The illustration of butterfly wing flaps in Brazil causing a tornado in Texas three weeks later illustrates **amplification**—a small change rapidly growing into a large effect. A butterfly wing flap disturbs a small number of air molecules. For this tiny disturbance to amplify into a tornado thousands of miles away suggests an exquisite sensitivity to such minute disturbances. Scientists call this **sensitive dependence**, a property of a system such that the smallest of changes now can rapidly produce very large effects in the system's future behavior.

But how sensitive are we talking about? Consider a car at a stop sign. When it is your turn to go, if you press the accelerator a little bit, the car inches forward slowly. Press the accelerator a little more and the car increases speed a little more. The small changes in the accelerator do not lead to rapid increases in the car's speed. This is not the sensitive dependence characteristic of chaos.

Consider another example. Suppose a perfectly symmetric cone is precisely balanced on its tip with only the force of gravity acting on it. Absent any other influences, the cone would maintain this unstable equilibrium forever. In the actual world, the perfect balance is unstable because the smallest nudge from an air molecule colliding with the cone will cause it to tip over. However, the cone could tip over in any direction due to the slight differences in various perturbations arising from suffering different collisions with different molecules.

This example illustrates that variations in the slightest of causes produce dramatically different effects. If we plotted the tipping over of the unstable cone, we would see that from a small ball of potential starting conditions representing the variations in how the air molecules strike the cone (apparent randomness), several different directions for the cone's falling issue forth from this small ball of uncertainty. Dramatic, indeed, but not what scientists typically mean by the sensitive dependence of chaotic dynamics.

Mathematical maps and checkerboards

One of the marks of chaos scientists look for is sensitive dependence leading to **exponential growth** in uncertainties. Think about the small ball of initial conditions representing variations in the starting point of a system. This ball represents the initial uncertainty in the measurement of the initial state or conditions of the system. In other words, our measurement of the initial conditions is somewhere within this small ball. This is why you often see scientists reporting measurements with error bars, a + and - telling us the range of uncertainty within which a measured value lies.

Suppose the uncertainty grows linearly with time. Think of linear growth in the following way: Imagine a checkerboard. Place one penny on the first square, two pennies on the second square, three pennies on the third square, and so forth. We will end up with 64 pennies stacked on the last square. This is a rule for adding the same number of pennies—one—to the previous number of pennies. A mathematical rule for generating a number from another number is called a **map**. The total number of pennies on the checkerboard will be 2080. This would represent the solid line in Figure 1.1.

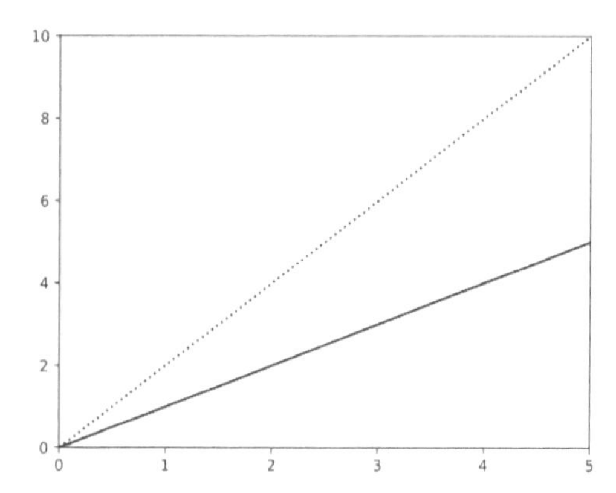

Figure 1.1. Uncertainty in initial conditions growing linearly with time.

Suppose we change the rule to always add two pennies to the previous number: two pennies on the first square, four pennies on the second square, six pennies on the third square and so forth. By the time we reach the 64th square, the total number of pennies on the checkerboard would be 4160. This map would be represented by the dotted line in Figure 1.1. The dotted line increases more quickly because it has a higher **slope**—the rise in the line along the vertical axis over the run along the horizontal axis. As the slope is increased—the number of pennies added to the previous number—the line would become increasingly vertical.

Contrast this with exponential growth. Exponential growth outpaces linear growth very rapidly (Figure 1.2). Continuing with our checkerboard analogy, again start with placing 1 penny on the first square. However, on the second square about 2.7 pennies are stacked, on the third square about 7.4 pennies are stacked, and so forth. On the last square we end up with about 6×10^{27} pennies. That's 6 followed by 27 zeros! In comparison with our linear maps in Figure 1.1, even for the steeper dotted line at square four there are eight pennies; by contrast for our exponential map at square 4 there are about 54.6 pennies.

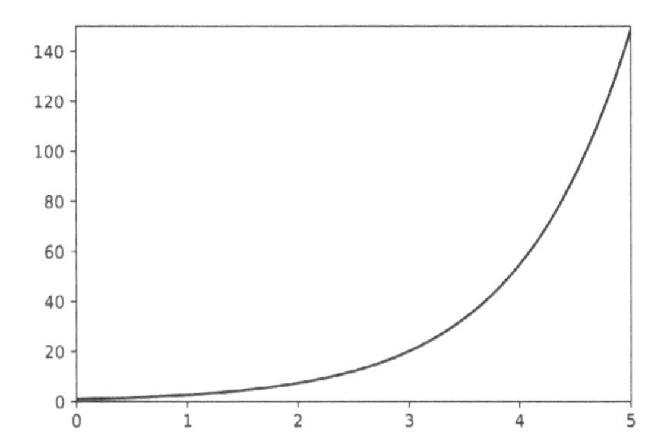

Figure 1.2. Uncertainty in initial conditions growing exponentially with time.

Why the value of 2.7 for the second square you might ask? Because exponential growth is characterized by the exponential function, $e^{\lambda t}$, where λ is a parameter. This is a map that tells us how the number of pennies grows with each square exponentially. In our example, $\lambda = 1$, and t is 0 for the first square, 1 for the second square, 2 for the third square and so forth. When time has a discrete value such as 1, 2, 3…, we say that we have a **discrete map**. Mathematically, $e^0=1$, e^1, is about 2.718, e^2 is about 7.389, and so forth. For the exponential map, we see that its value grows rapidly with increasing t as illustrated by the checkerboard analogy and by Figure 1.2.

Geometric growth rates can also be very rapid, but we will not bother with those. Now you are seeing part of what is exciting about chaos! Sensitive dependence implies that the growth rate in any small changes to the initial state of the system can lead to large differences in system behavior very rapidly. Put another way, any uncertainty in the initial state of a system very quickly can lead to very large uncertainty about the system's behavior when the uncertainty grows exponentially with time.

For instance, suppose you have some number of pennies as the initial condition but do not know the exact number. Consider our linear map and suppose you can reduce the uncertainty in the measurement of the initial number of pennies by a factor of one hundred. You then could forecast the growth in the number of pennies one hundred times longer into the future before the growth in uncertainty exceeds this factor of one hundred. By contrast, with the exponential map, reducing the initial uncertainty by one hundred will only allow you to accurately forecast the future growth in the number of pennies four times as long into the future. This represents a significant limitation on the future forecast if the model exhibits exponential relative to linear growth.

Typically, it is not possible to reduce the uncertainty to zero in our measurement of the initial conditions due to mechanical and other limitations in our observations. There are no perfect measurements. Hence, sensitive dependence in our models has actual world consequences for our forecasting the future for systems behaving chaotically.

Take weather forecasting as an example. The physicist and mathematician Henri Poincaré (1854 – 1912) noted that cyclone behavior was sensitive to the precise location where they originate:

> We see that great perturbations generally happen in regions where the atmosphere is in unstable equilibrium. The meteorologists are aware that this equilibrium is unstable, that a cyclone is arising somewhere; but where they can not tell; one-tenth of a degree more or less at any point, and the cyclone bursts here and not there, and spreads its ravages over countries it would have spared. This we could have foreseen if we had known that tenth of a degree, but the observations were neither sufficiently close nor sufficiently precise... Here again we find the same contrast between a very slight cause, unappreciable to the observer, and important effects, which are sometimes tremendous disasters (1921, p. 398).

Note the intersection between apparent randomness and sensitive dependence. Because the observations of the origin point are not precise enough, cyclones burst on the scene with their ravages in a manner that appears random. Although Poincaré was not aware of the modern notion of sensitive dependence, the uncertainty in the precise origin of cyclones is the same as the case where a relatively small ball of uncertainties grows exponentially rapidly.

Modern meteorology, though greatly improved in forecasting skill, still must contend with sensitive dependence due to chaotic dynamics and uncertainty in our data (more on this later). Figure 1.3 shows a computer forecast for tropical storm Elsa generated on July 1st, 2021, before it became a hurricane. Because there is some imprecision in knowing the precise starting conditions for Elsa's origin, the uncertainty in the actual path the storm will take spreads rapidly. This spread in the possible paths the storm will take is represented by a cone.

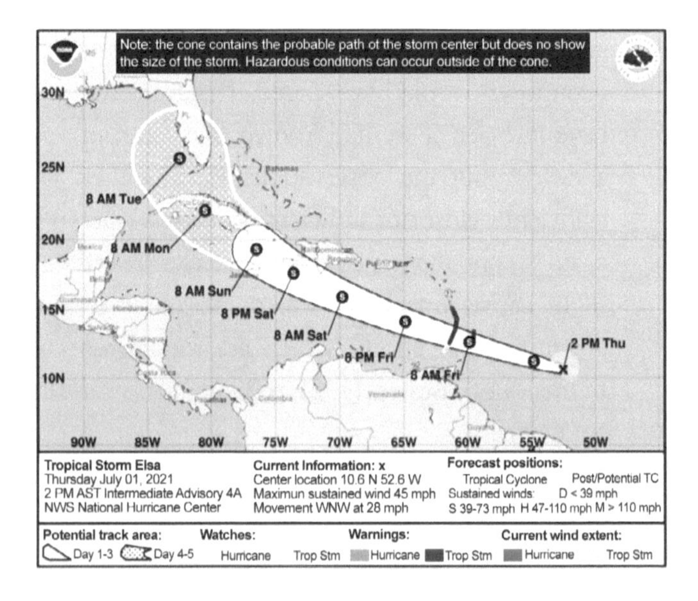

Figure 1.3. Computer forecast for tropical storm Elsa. Given the imprecision in the initial starting conditions for Elsa, the uncertainty spreads rapidly leading to a large cone of possible paths the storm could take. Courtesy of the National Oceanic and Atmospheric Administration (NOAA).

The forecast in Figure 1.3 applies techniques to help mitigate the challenge presented by the growth in the uncertainty in the initial conditions (more on these later). Absent these techniques, the spreading of the cone in Figure 1.3 would be even larger.

Scientists are helped by another fact when making forecasts under uncertainty with exponential growth in those uncertainties—such growth does not go on forever. The finitude of physical space or other resources means that growth in uncertainty will saturate at a finite amount. Return to our example of exponential growth in pennies on the checkerboard. Restricting ourselves to US pennies, it is estimated that about 200 billion are currently in circulation. This means that the growth in the number of pennies would stop at approximately square 21. Even if we added in all the British pence and other pennies from all other countries, there are not enough pennies in the world to support growth to square 22.

Edward Lorenz and the "birth" of chaos

Because weather forecasts have important agricultural, economic, and health consequences, significant attention has been paid to developing accurate weather forecasting models over the decades. In fact, weather models were some of the first applications for the earliest computers. There has always been interest in improving the accuracy of forecasting models. Because of debates over different approaches to forecast modeling, Edward Lorenz (1917 – 2008), a mathematician who spent his career at the Massachusetts Institute of Technology, decided to put the two main approaches to a test.

It was in this context that he developed a simplified weather forecasting model and sought to find a set of **aperiodic** solutions. Periodic solutions repeat a pattern such as the second hand of a clock returning to 12 every 60 seconds. Aperiodic solutions do not repeat such a pattern (at least on reasonable timescales; demonstrating a solution never repeats is very difficult). It turns out that the forecasting approaches could not be distinguished using periodic solutions. He first tried a 12-equation model, but later with the help of a colleague tried out a much simpler set of three equations.

Today, the program for such models would be developed and run on a desktop or laptop computer, a workstation, or even a high-performance computer cluster. In 1960, such things did not exist. Instead, the computer Lorenz's model ran on was the size of a desk and so loud it had to be put in its own office. The actual computer programs were developed and run by Margaret Hamilton and Ellen Fetter, which required creating the binary code on long spools of paper tape (Sokol 2019). This was a very tedious process requiring a great deal of mathematical and computer science skill that we usually take for granted when telling the story of chaos' birth as a subject of study.

After some trial and error, Lorenz found parameter settings for his model that treated the heat in the atmosphere simply and transparently (though not realistically), which produced aperiodic solutions. He was able to show that the solutions would not repeat themselves because he discovered that they had a deep relationship to the Cantor set (see Chapter 6). Lorenz demonstrated that the two main modeling approaches were strongly distinguished when solutions were aperiodic. This served as proof of principle for his claim that the results of the dynamical modeling approach could not be reproduced by the so-called linear approach (more on what linear means below).

But there was something peculiar about his numerical model. Lorenz decided to rerun the model with some of the model's own output from an earlier time to dive into more detail about how the weather in the model was behaving. What he noticed was that even using the model's own data from an earlier time—the equivalent to backing the model weather up several weeks to run again—the computer output began to diverge from the original run.

Lorenz and his team discovered that the culprit was in the computer's data that fed back into the model. The computer had been keeping track of data to six decimal places while the printout contained data rounded to the third decimal place to save paper. But it was the data from the computer printout that was fed back into the computer, so these were not the exact values that the computer had calculated previously. The small difference in rounding the input had been amplified rapidly by the computer model. Lorenz had found chaos. Though the phenomenon had yet to receive that name. The earliest use of the term "chaotic" to describe the phenomenon Lorenz observed was in a published paper by David Ruelle and Floris Takens (1971). (The term "strange attractor"—see below—also first appears in this paper.) However, it was the influential paper by Tien-Yien Li and James A. Yorke published in 1975 with the title "Period Three Implies Chaos" that led to the widespread use of the term "chaos" for these mathematical behaviors.

To illustrate the phenomenon, Lorenz adopted a simpler set of model equations that captured some features of fluid convection in the atmosphere and studied the chaotic dynamics (1963). Lorenz's paper was one of the foundational papers kicking off the study of chaotic dynamics. There is a case to be made that he "discovered" it, though, as we can see from Poincaré's remarks above, some kind of awareness of the phenomenon has been around for a much longer time. And mathematicians who have been interested in the stability of solutions to equations, such as Poincaré, Aleksandr Lyapunov, Philip Franklin, and Andrey A. Markov, also discussed conditions for stability and proved when solutions were stable. When solutions failed to meet these stability conditions, the implications for chaotic dynamics were there, but remained largely unrecognized and unexplored.

One lesson learned from sensitive dependence relevant to weather and other kinds of forecasting: Collect data that is as accurate as is possible. The constraints on accuracy typically are precision of measurement instruments and money. It costs money to have the most accurate instruments possible positioned where they do the most good generating useful initial conditions to feed into our models. Investments in our measurements have paid dividends for the kinds of forecasting modern societies depend upon.

Likewise, it costs money to develop and run sophisticated forecasting models on supercomputers. Hence, meteorologists usually can only run a sample of initial conditions to form their ensemble forecasts. Given limited resources, scientists end up balancing the costs of improving measurements versus the number of model runs on supercomputers that can produce useful ensemble forecasts.

Chapter 2

Nonlinearity

Where does the sensitive dependence of our models come from? It arises from a property of models known as **nonlinearity**. If a model is linear, any multiplicative change in a variable, by a factor α say, implies a proportional change of its output by α. Think of your stereo at low volume. If you turn the volume control one unit, the volume increases by one unit. If you now turn the control two units, the volume increases two units. This is an example of a linear response. Our maps for adding pennies, illustrated in Figure 1.1, are examples of linear maps.

In a nonlinear model, such as the one Lorenz examined, the output need not be proportional to the change in α. If you turn your stereo volume control too far, the volume may not only increase more than the number of units of the turn, but whistles and various other distortions occur in the sound. This illustrates a nonlinear response. The exponential map illustrated in Figure 1.2 is an example of a nonlinear map. There are technical conditions determining when a mathematical model is nonlinear, but these need not occupy us presently.

The effects of nonlinearity lead to the aperiodicity of solutions that exhibit sensitive dependence. Roughly, whenever the nonlinearities in a model begin to dominate the model behavior, sensitive dependence comes into play. However, triggering the nonlinearities can be somewhat tricky. Consider the **damped driven pendulum**. As an example, think of a playground swing. The swing is a pendulum with an anchor point for the chains at the top. The length of the chains hangs down, attaching to the seat where a person sits. If you give the swing a push it will oscillate back and forth, but the amount of back-and-forth motion will decrease with each oscillation due to friction at the anchor point plus some air friction. Eventually the swing comes to a halt. This would be an example of a damped pendulum. Now suppose that you keep pushing the swing at the completion of each oscillation providing energy to keep the swing going. This would be an example of a damped driven pendulum where your pushes are the driving force on the pendulum.

The length of the pendulum is integral to the pendulum's natural frequency of oscillation. In other words, setting the pendulum in motion with no damping—no friction—it would oscillate back and forth at a frequency determined by the pendulum's length. For the damped driven pendulum, there are additional factors: the magnitude of the force driving the pendulum, the frequency of the driving (the frequency of pushing), and the magnitude of the friction damping the pendulum motion. For many combinations of these three factors, the damped driven pendulum never exhibits chaotic behavior. It acts like a linear system with small changes to the initial conditions leading to only small changes in the pendulum's behavior.

Yet, for some combinations of driving force magnitude and frequency, and magnitude of damping, the pendulum exhibits chaotic dynamics. For these combinations of factors, any small change in the initial conditions for the start of pendulum motion yields rapid large changes in the behavior of the pendulum from a slightly different initial condition. We have sensitive dependence. One can find the different combinations of factors leading to chaotic behavior by examining the mathematical model for the damped driven pendulum.

This example illustrates that although chaotic dynamics is due to nonlinearity, the parameter values of the model are crucial to whether a system exhibits chaotic behavior or not. In other words, nonlinear models are not chaotic all the time but only under particular combinations of parameter values. This circumstance suggests digging deeper to understand what is special about the nonlinear dynamics leading to aperiodicity in our mathematical models, which we will return to.

Chapter 3

Dynamical Systems and Determinism

Chaotic dynamics is typically understood as a mathematical property of a **dynamical system**. A dynamical system is a deterministic mathematical model that can be thought of as a description or rule for how the observable properties of a system evolve in time. Examples can be models of anything from the number of pennies growing over time, to the temperature readings at a weather station, to the appearance of sunspots, to the orbiting of the planets. Such models may be studied as mathematical objects or may be used to describe target systems to be studied such as physical, biological, or economic systems.

A dynamical system's mathematical rule specifies how an initial number (or set of numbers) input into the model will produce a new number (or set of numbers) as output. This output is then used as the new input so that the rule produces yet another new number (or set of numbers). This process is called **iteration**, where the model's output forms the new input for the model. The output at each iteration produces a **time series**, a list of numbers (or sets of numbers) at each time ordered from earlier to later. The exponential map for growth of pennies on the checkerboard is an example of a dynamical system with the output for each square representing the value in a series in time (where each square is a time step). Or think of the moment-by-moment rise and fall of a stock price or the daily infection counts from Covid-19 displayed on a graph.

Mathematicians call a one-dimensional dynamical system a map. Figures 1.1 and 1.2 are examples of maps you have already met. For two or higher dimensions, mathematicians call dynamical systems **flows**. Think of the three-dimensional flow of water from your faucet. Figure 1.3 is an example of a flow.

Another way to think about a dynamical system is as a mathematical rule for how a system behaves in **state space**. A state space is an abstract mathematical space of points where each point represents a possible state of the system. A state is characterized by the instantaneous values of the independent variables considered crucial for a complete description of the state. The number of independent variables needed to completely characterize the state determines the number of dimensions of the state space. The initial state is the first number we feed into our mathematical map. Iterating our map beginning with this initial state produces a **trajectory** in state space, a series of state transitions in time from one state to another to another for as long as the iteration process continues. Such a trajectory represents a time series of values in state space.

Along with a mathematical rule and an initial state, a dynamical system also has one or more **parameters**, factors that characterize how strong or weak a particular term in the mathematical model contributes to the dynamical system's behavior. As you will see, all three of these features—the rule, initial state, and parameters—are important to the appearance of chaotic behavior.

The aperiodic solutions Lorenz discovered were features of a dynamical system. He hit upon a combination of model, initial state, and parameter values that produced chaotic behavior for a nonlinear system. The study of chaos has been carried out for a variety of such mathematical systems. This is one reason why chaotic dynamics is usually understood to be a property of deterministic models. Another reason is that chaotic behavior looks to be ruled out for quantum systems (Chapter 10), and quantum mechanics is usually held up as the quintessential example of indeterminism. Think of the irreducible randomness described in the Introduction.

Deterministic models

But what makes a model deterministic? Consider a movie. Start the movie over and over at the same frame (analogous to starting the model with the same initial state), then the movie repeats every detail of its total history every time it plays. Moreover, identical copies would produce the same sequence of events. No matter whether you always start *Jurassic Park* at the beginning frame, the middle frame, or any other frame, the same sequence of frames plays out. The T-rex as antihero always saves the day. No new frames are added to the movie, and none disappear.

By way of contrast, imagine a device that generates a different sequence of pictures on some occasions when starting from the same initial picture. Further imagine that this device has the property that simply by choosing to start with any picture normally appearing in the sequence, sometimes the chosen picture is not followed by the usual sequence of pictures. Or sometimes a picture does not appear in the sequence, or that new ones are added. A model behaving like this would be nondeterministic.

To put this in terms of dynamical systems and time series, starting a dynamical system off with the same exact initial value and the same time series as the output of iterating it, then the dynamical system is deterministic (like a movie). In contrast, if starting off a dynamical system with the same exact initial value leads to different time series being generated each time the process is repeated, then the dynamical system is nondeterministic (like the device generating different sequences of pictures even when starting with the same initial picture).

A mathematical model exhibits **unique evolution** if a given state of a model is always followed by the same history of state transitions (like *Jurassic Park*). Unique evolution is the core property that makes a model deterministic (Bishop 2005). Some models exhibiting unique evolution are predictable. For instance, given the same initial starting conditions, a mathematical model of a pendulum will run through the same history of state transitions in a predictable way like a movie. But not all deterministic models are predictable, at least in all aspects. For example, any models exhibiting chaotic dynamics will be deterministic, but some of their behaviors are unpredictable at least after a short time (more on this later). The exact trajectory of a hurricane in the Atlantic can be unpredictable very far into the future because of chaotic dynamics even though it is a deterministic system (Figure 1.3).

As you saw in Chapter 1, there is a relationship between aperiodic solutions and sensitive dependence. Mathematical models exhibiting chaotic dynamics will behave according to the strictures of unique evolution only if started with the identical initial conditions. The surprising thing is that any small uncertainty in these initial conditions will be rapidly amplified by the model over time so that nearby solutions in state space will diverge away from each other quickly. Hence, the need for at least the properties of sensitive dependence and unique evolution for a mathematical model to be chaotic. This has implications for quantum mechanics as you will see later.

As a final comment, mathematicians like to be able to prove results in mathematics that always hold true. Physicists and others might aspire to this same goal for their work, but often must settle for something less than proving a result. Determinism is a quality like this. We usually cannot prove that a mathematical model has unique evolution for all times, but only for some finite range of time (Arnold 1988).

Chapter 4

Chaotic Dynamics: Complex Order

To say more about chaotic dynamics studied in dynamical systems, you need to know more about some basic properties of mathematical maps. One important property is that of an **attractor**. An attractor is a value (or set of values in the case of a flow) in state space that trajectories converge onto. The sense of converge onto for a map can be that the map gives the same value (or set of values in the case of a flow) repeatedly, or that the output of the map gets arbitrarily close to a specific value (or set of values in the case of a flow).

There are four different kinds of attractors:

1. Fixed points: when a dynamical system starts producing the same output on every iteration.
2. Periodic loops: when a dynamical system starts periodically repeating the same value (think of the big hand on a clock pointing to 12 every twelve hours).
3. Almost periodic loops: when a dynamical system has more than one period but does not repeat the exact values, only coming very close, yet there is a discernible pattern (think of the tides rolling in and out).
4. Aperiodic: when a dynamical system's output appears to randomly jump around but intricate order can be teased out (the chaos we seek lurks here).

These attractors will be important as the discussion moves forward, but when the outputs of a dynamical system contract or shrink down to some attractor, mathematicians say that it is **dissipative**. A dissipative dynamical system has the property of shrinking its activity to a smaller area or volume of state space. Observe this behavior, and we know that an attractor exists in that portion of state space. A dynamical system where the activity does not shrink to a smaller area or volume in state space is called **Hamiltonian**. Such systems are also known as **conservative** because they conserve the state space volume of the activity. Chaotic dynamics happen in Hamiltonian systems, too (Chapter 7).

In a seminal 1976 article published in *Nature*, physicist turned ecologist Robert M. May collected in one place the variety of complexity and order of mathematical chaos. He began his review with an ecological example: a seasonal breeding population where the generations do not overlap. Think of a crop pest that produces a new generation every season that a crop is growing but doesn't breed year-round. It is useful to know how the size of this year's pest generation is related to the size of last year's population for predicting future population size. The mathematical relationship between this year's population and last year's turns out to be nonlinear (Chapter 2). Likewise, the size of this year's population is directly related to last year's in a way that often can be characterized by a deterministic model (Chapter 3).

The behavior of this mathematical model turns out to be surprising. First, when a previous generation's population is small, population size increases in subsequent generations. Mathematicians call a function with such steady increase (or steady decrease) a **monotonic** function. Figures 1.1 and 1.2 are examples of monotonically increasing functions. However, when a previous generation's population is large, population size decreases monotonically in subsequent generations. Under many conditions, then, the model for population growth either increases strictly monotonically or decreases strictly monotonically generation by generation.

The simplest mathematical model having this behavior is the following. Let x_t represent the population of the current generation, where t represents last year. Then the size of the next population at time $t + 1$, this year, is related to the size of the population at time t by the following rule: Subtract x_t^2 from x_t and multiply this difference by a factor α. Mathematically, this rule is written as

$$x_{t+1}=\alpha x_t (1-x_t),$$

with α as a parameter whose value ranges from one to four and represents the density of resources available to the population. This equation is known as the **logistic equation** or **logistic map**. Figure 4.1 illustrates the behavior of the map where a small population x_t leads to a larger population in the next generation x_{t+1} but when the population x_t is too large the next generation x_{t+1} will decrease.

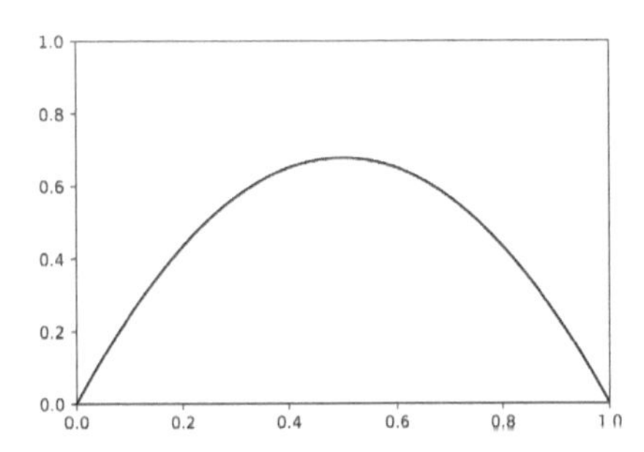

Figure 4.1. The logistic map plotted for α=2.707. The map rises to a maximum at x=0.5.

The logistic map is an example of a nonlinear dynamical system. It is nonlinear because it does not involve the variable x by itself—anytime state variables are multiplied by state variables the mathematical map will be nonlinear even if it is one-dimensional, as in this case.

Furthermore, we can prove that the logistic map has an attractor for all values of α less than four. You can see this from looking at Figure 4.1: The largest value x can take on is $\alpha/4$ since all values are confined to the range from zero to one. Therefore, this map must have an attractor under this condition. For small values of α the attractor is zero. For any value of x greater than one, values of x greater than zero all move away from zero. As you will see, the characteristics of the attractor depend on the value of α. For a population model, this makes sense. When a population is small, it underuses resources leaving more for the next generation. In contrast, when a population is too large, it overuses resources leaving less for the next generation.

Stability

Suppose α=0 and let the initial value of x_t, call it x_0, be 0.5. The first iterate of the logistic map will be x_1=1.0. The next iterate—the next step in time—is x_2=0. But now, as you can see from the form of the logistic map, all further iterates will yield zero. Hence, the time series for this situation, beginning with t=0, is 0.5, 1, 0, 0, 0…This is an example of a fixed-point attractor. For such an attractor, the uncertainty in the initial conditions will cease growing as the trajectories in state space converge to the attractor.

A natural question to ask about an attractor is how stable it is—will it persist and under what circumstances? One can show mathematically for the logistic map that the fixed-point attractor is stable for all values of α between one and three. For example, we can plot the stability of the fixed point for α=2.707 (compare Figure 4.2 with Figure 4.1) and see that the logistic map rapidly converges to returning the same value.

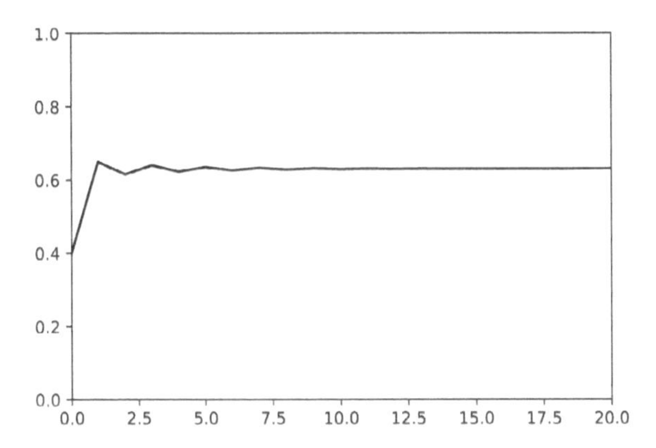

Figure 4.2. Stability of the logistic map for α=2.707. After a few time steps, the map settles into returning the same output for all subsequent iterations.

What happens when α is greater than three? Suppose we let α=2.9.

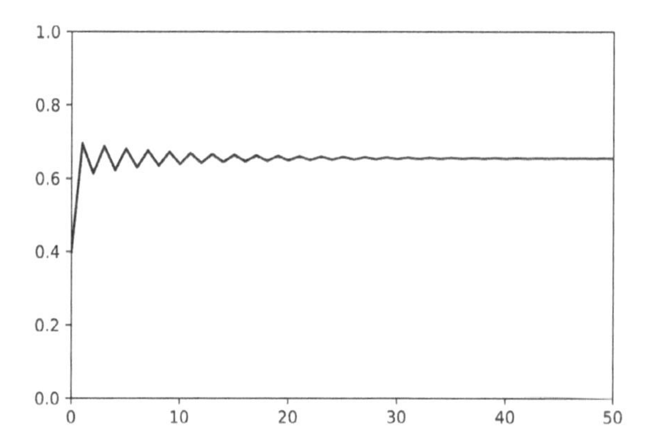

Figure 4.3. Stability of the logistic map for α=2.9. The map takes much longer to settle down to returning the same output as the value of α approaches three.

Comparing Figures 4.2 and 4.3, observe that the logistic map takes much longer to converge to the point attractor. The closer α approaches three, the longer the map takes to settle down to the attractor. At α=3, there is a **bifurcation** where the logistic map settles into a period loop attractor of period two meaning the map settles into returning alternating outputs every time step. For instance, consider α=3.01 (Figure 4.4). Keep in mind that the fixed-point attractor has not vanished; rather, it is now an unstable fixed point. For α=3 or greater, any trajectory of the logistic map that approaches the fixed-point attractor will not settle down there to repeat the same output value for every subsequent iteration. It will move on to follow one branch or the other of the bifurcation.

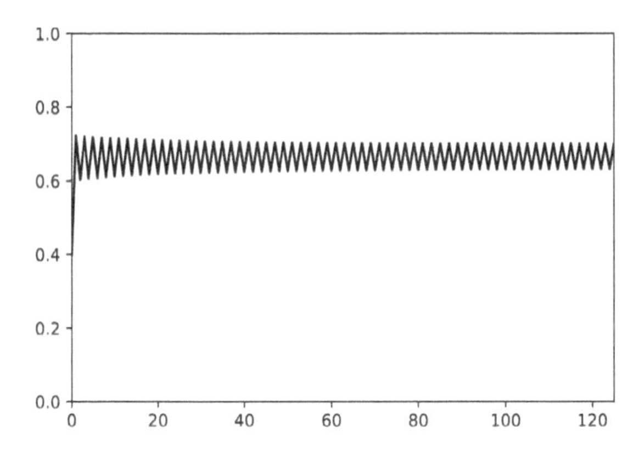

Figure 4.4. For α=3.01, the logistic map has a period loop attractor that repeats the output value every second time step (period two).

When α=3.5 another bifurcation occurs, and the logistic map settles into a period four attractor. Continuing to increase α leads to successive bifurcations where the map settles into attractors of period eight, then 16, then 32, then 64, and so forth. In other words, the attractor period doubles with every bifurcation. Figure 4.5 illustrates several of these period doublings: periods two, four, eight and 16 for corresponding values of α. And as with the fixed-point attractor, the period two and successive attractors do not vanish. Instead, they become unstable and trajectories approaching them will follow one or another branching path of the new loop attractor.

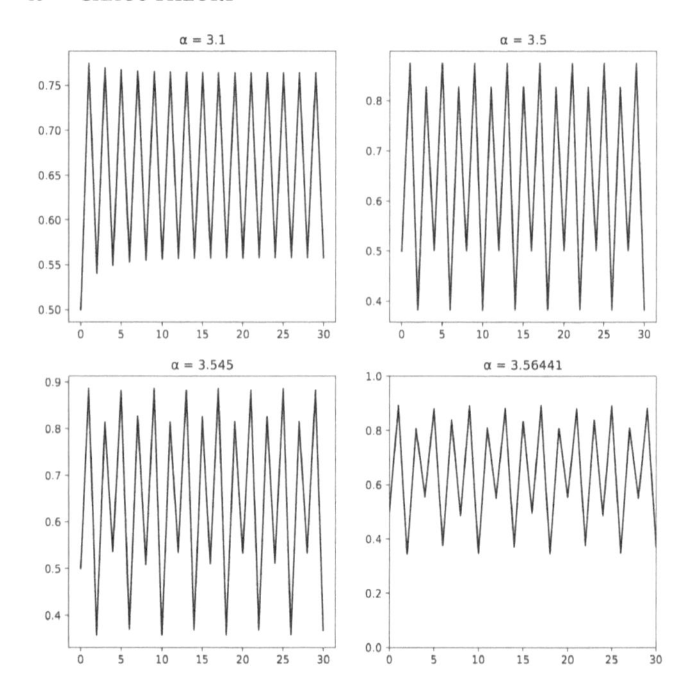

Figure 4.5. A sequence of plots of the logistic map showing period doubling of the attractor: period two (α=3.1), period four (α=3.5), period eight (α=3.545), and period 16 (α=3.56441).

Chaos

Notice that the values of α get closer together for each bifurcation. Clearly something interesting is going on as α continues to increase. For α=3.7 something remarkable happens, illustrated in Figure 4.6.

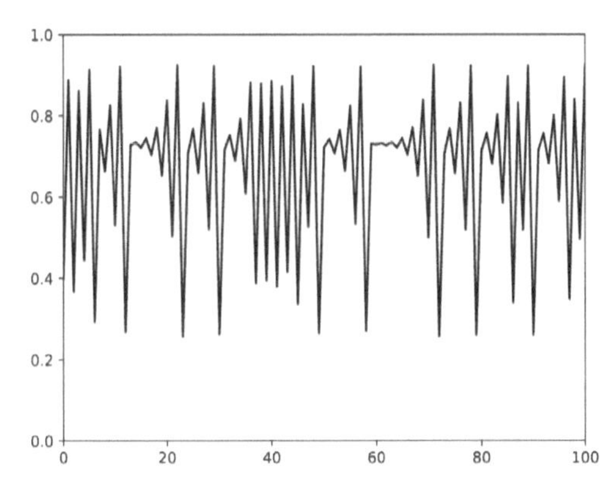

Figure 4.6. For α=3.7, the logistic map no longer has an attractor. Instead, its output is now aperiodic

The outputs for successive iterations no longer repeat with any period that we can discern. In other words, the logistic map now yields aperiodic outputs. We have entered the realm of chaotic dynamics for the logistic map. Even though the behavior appears random, it is still deterministic (apparent randomness) because it is generated by a deterministic equation.

Now the beauty of the hidden order of chaos also comes into view. Suppose we compare the logistic map for α=3.7 but change the initial condition by one ten thousandth. Figure 4.7 shows the result. The outputs of the logistic map—the two trajectories starting with slightly different initial conditions—look the same for a while but begin to diverge from each other at about time step 50. Trying this same exercise with any of the values of α for any of the attractors illustrated in Figures 4.2-4.5 will yield no difference in trajectories whatsoever. We have sensitive dependence on the initial conditions, a hallmark of chaos.

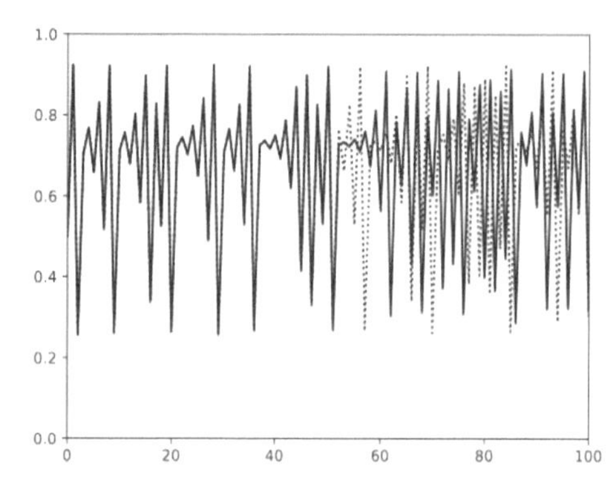

Figure 4.7. Comparison of the logistic map for α=3.7 but with two initial conditions that vary by one ten thousandth.

As pointed out earlier, the behavior of the logistic map is determined by the value of α. You have seen some of this in the previous figures in this chapter. A particularly impressive way of seeing this is by plotting the bifurcation points in a **bifurcation diagram**.

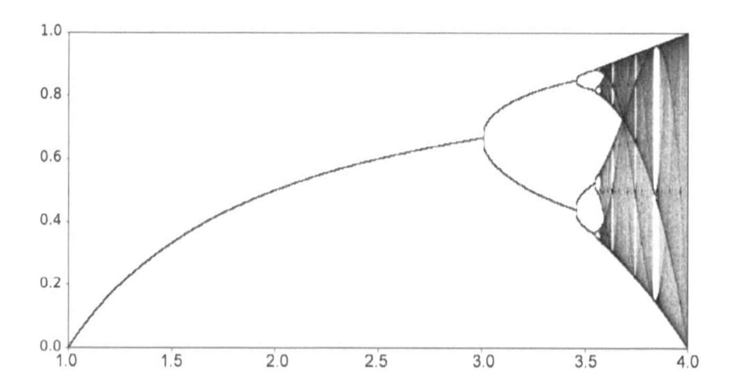

Figure 4.8. Bifurcation diagram showing the sequence of period doublings for the logistic map as α increases. Eventually, period doubling leads to chaos.

In Figure 4.8 you can see the sequence of period doubling noted earlier. These look like pitchforks and are called **pitchfork bifurcations**. As you scan from left to right, α increases while the output of the logistic map is plotted for that value of α. For α up to three, you can see the fixed-point attractor that the logistic map converges to. At $\alpha=3$ period doubling occurs. Continuing from left to right, you can see another period doubling event leading to a period four attractor, then period eight, then period 16.

As α continues to increase, the period doublings come very rapidly and eventually make a transition to period three attractors starting at period three and then bifurcating to six, then to 12 and so on. These period three bifurcations become increasingly rapid as α increases until period four bifurcations start, then period five and so on through the integers until the plot begins to look like a jumble. This "jumble" is the chaotic behavior of the logistic map revealing an exquisite sensitive dependence on the value of α.

At first glance, the right-hand side of Figure 4.8 looks random. Yet, this dense jumble is filled with intricate order as Figure 4.9 reveals. Zooming in on the plot for α between 3.7 and 3.9, you can see that there are regions where there appears to be random behavior in the logistic map. This is where the aperiodic trajectories are such that the logistic map never repeats any of its output values exactly, so far as we can determine. Nevertheless, notice that this behavior is punctuated by regions where period loop attractors reemerge of every integer period. For instance, you can see period doubling sequences emerge once again, and at α=3.83 a period three orbit emerges. If you look closely, you can find loop attractors of five and seven.

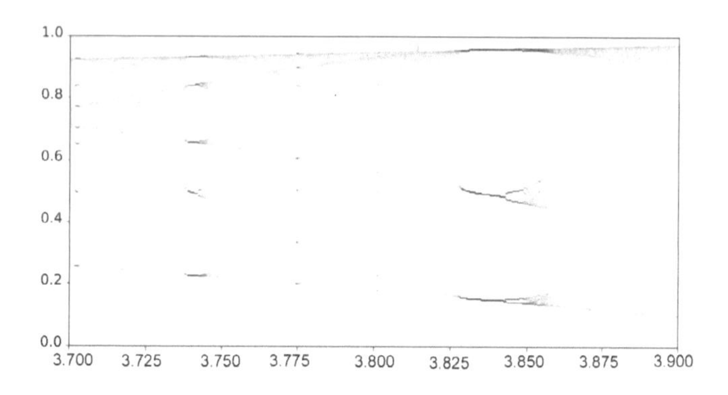

Figure 4.9. Zooming into bifurcation diagram of the logistic map for values of between 3.7 and 3.9 reveals intricate order characteristic of chaotic dynamics.

There is a surprising regularity hidden from view in Figures 4.8 and 4.9: Consider any three consecutive values of α you like at which there is a period doubling. Subtract the first value from the second. Divide this result by the difference between the second and third values. What physicist Mitchell Feigenbaum (1944 – 2019) discovered in the late 1970s was that this procedure will always yield –4.6692016091, the **Feigenbaum constant**. This regularity means that each succeeding value of α where a doubling bifurcation takes place gets closer together.

Here is another feature of chaos that may surprise you. Figures 4.8 and 4.9 suggest that there are lots of value of α for which the logistic map behaves chaotically. And you would be right. In fact, there is an infinite number of such values. But when compared to the total number of values of α for which there are stable attractors, the number of values for chaotic behavior is of **measure zero** as the mathematicians say. This roughly means that although there are infinitely many values of α leading to chaotic behavior, there is a much larger infinity of values leading to attractors. The set of values leading to chaotic behavior is easily contained within the larger set of values of α for which the logistic map's behavior is always nonchaotic. The bottom line is that there are far more values of α leading to stable behavior than not. Although true for the logistic map, this result is true for all chaotic maps and flows in general: There are always vastly more parameter values leading to stable than to chaotic behavior.

The intricate order I have been describing in this chapter is characteristic of chaotic dynamics, not lawless randomness. And there are more surprises in store! But for the moment, note two things illustrated by the logistic map: (1) A very simple mathematical model can produce very complicated behavior. One does not need complicated models to study complex behavior. (2) This simple model's chaotic dynamics exhibits sensitivity to the values of x and α. It is this sensitive dependence that has generated much of the excitement about chaos. The value of the parameter α determines when sensitive dependence behavior of the logistic map kicks in. When it kicks in, even the smallest changes in values of x yield dramatic changes in the behaviors of the system.

Chapter 5

Chaotic Dynamics: Sensitive Dependence Again

The concept of sensitive dependence was introduced in Chapter 1 with the exponential map being the prime exemplar (Figure 1.2). Exponential growth in uncertainties has long been considered a defining characteristic of chaotic dynamics. Mathematicians make this characterization more precise using the **Lyapunov exponent**, named after Russian mathematician Aleksandr Mikhailovich Lyapunov (1857 – 1918). Recall that the exponential map was introduce as the function $e^{\lambda t}$. In chaos studies, the parameter λ is the Lyapunov exponent.

What does the Lyapunov exponent mean? It is a parameter characterizing the average rate of divergence of any uncertainty in the initial conditions for a chaotic system. This average rate is calculated or estimated over many initial conditions. In other words, by tracking the rate at which neighboring trajectories issuing forth from the small ball of initial conditions diverge from each other if the value of λ is positive (e.g., see Figure 4.6). If λ is negative, then nearby trajectories will converge towards each other. Chaos is characterized by positive Lyapunov exponents.

Another way to think of the Lyapunov exponent is as giving the stretching rate per iteration of the map averaged over a trajectory. Stretching in this context means the rapid divergence of neighboring trajectories away from each other. This stretching is related to the presence of nonlinearities in the mathematical model. Figure 5.1 shows the values of the Lyapunov exponents for the logistic map compared to the bifurcation diagram.

Comparing the two diagrams shows that the sign of the Lyapunov exponent tracks well with the bifurcation behavior. Indeed, for mathematical models such as the logistic map, we always find this tracking to be the case. For instance, compare the region of Figure 5.1 where the values of α vary between 3.7 and 3.9. Notice that there are negative Lyapunov exponents in this region indicating that the logistic map has stable attractors of period two and so forth. Everywhere there is a positive Lyapunov exponent (indicated in grey), we have sensitive dependence on the initial values of x.

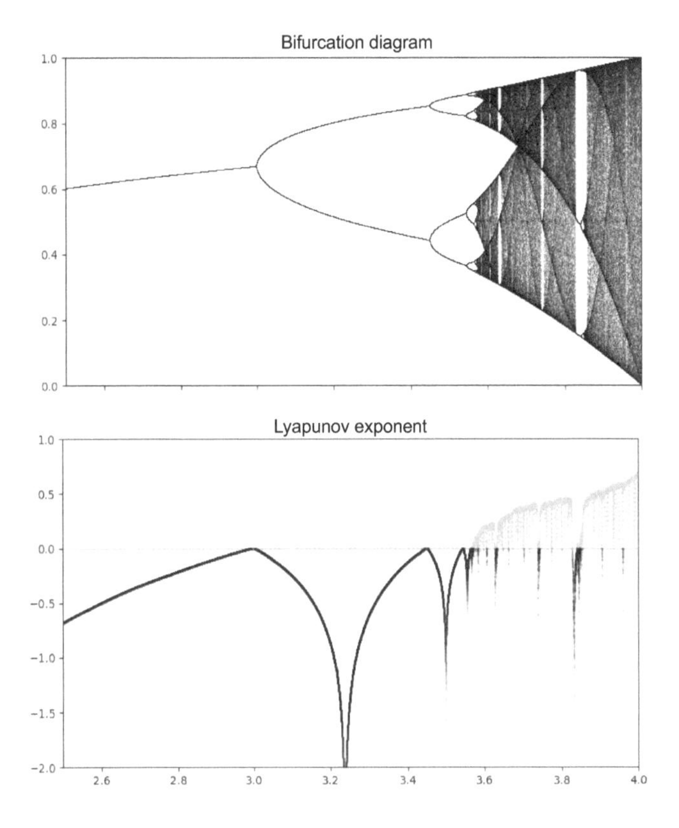

Figure 5.1. A plot of the values of the Lyapunov exponent for α ranging from 2.5 to 4.0 at the bottom with the corresponding bifurcation diagram at the top. For the Lyapunov exponent values, black is negative while grey is positive.

Things are more difficult when it comes to estimating Lyapunov exponents from time series data. Recall, that a time series is a list of numbers (or set of numbers) that are outputs ordered in time forming a trajectory in state space. When scientists make measurements of the dynamics of physical systems and societies (e.g., the daily closing value of the New York Stock Exchange), they generate a time series.

In principle, a scientist can track the behavior of these time series in state space and check to see if nearby trajectories diverge exponentially fast from each other on average by extracting an estimate for the Lyapunov exponents. And we have known for decades that the complete dynamics of a system can be constructed from a suitably obtained time series providing a basis for estimating Lyapunov exponents (Takens 1981), though this is more difficult than it sounds. One problem with producing these estimates is that our algorithms for doing so need to be able to find both positive and negative exponents. If an algorithm is tuned to only find positive Lyapunov exponents and we are studying a stable, nonchaotic system, our algorithm will never tell us we have a stable system.

Another complicating factor is that all observations are noisy, as noted several times. This means that the time series data contains uncertainties, and this noise tends to degrade the accuracy of our Lyapunov exponent estimates. Sophisticated mathematical techniques overcoming these problems have been developed (e.g., Yang and Wu 2011).

An advantage of the Lyapunov exponent is that it can serve as a convenient measure of the growth in the uncertainty and, hence, give some insight into system predictability. For a linear system, the uncertainty is usually characterized by the doubling time, 2^t. This is the time over which we expect the uncertainty to grow to twice that in our initial data. For chaotic systems such as the logistic map, the chaotic doubling time is $2^{\lambda t}$ because the growth in uncertainty is thought to be a pure exponential.

However, since most chaotic systems only exhibit on average exponential growth, things are more complicated. Such systems have as many Lyapunov exponents as there are directions trajectories can take in a state space. It is customary to rank the Lyapunov exponents in order from greatest to least. The largest is called the **leading Lyapunov exponent** and this is the exponent used in estimates of the growth of uncertainty in the time series of a chaotic system.

One might think that this gives scientists a good way of estimating how long into the future a system's behavior can be predictable, but there is a further subtlety. Because of how they are defined mathematically, Lyapunov exponents can only reflect the growth in uncertainty infinitesimally close to a trajectory. If the uncertainty only grows infinitesimally, this has no effect on our ability to predict system behavior. It turns out that the doubling time often is a more useful measure of predictability for chaotic systems than using Lyapunov exponents.

Problems defining chaos as the existence of a positive Lyapunov exponent

Nevertheless, there are problems with looking to a positive Lyapunov exponent as THE definition of chaotic dynamics. First, there is the infinitesimal nature of the growth characterized by such exponents just mentioned. Second, the procedure for defining Lyapunov exponents assumes an infinite time limit. This means that positive exponents only characterize growth as time tends to infinity. Since all our systems are temporally finite, defining chaos this way is inconsistent with the world we live in.

Moreover, Lyapunov exponents are more nuanced than the "on average" language indicates. Calculating finite time Lyapunov exponents at any moment in time does not lead to exponential growth in uncertainties (e.g., Smith, Ziehmann, and Fraedrich 1999). Finite-time Lyapunov exponents vary from point to point in a system's state space. This means that trajectories issuing forth from a small ball of uncertainty in state space diverge and converge from each other at various rates as they evolve in time. In other words, the uncertainty does not increase or decrease monotonically in the chaotic region of state space except for some simple mathematical models. Indeed, there are some locations on attractors, such as the one Lorenz discovered, where a small ball of uncertainty in the initial data will not grow at all with time. If we focused on such locations and computed Lyapunov exponents, we would miss the chaotic dynamics altogether!

Another issue is that uncertainties saturate at the size of whatever attractors are operative in a chaotic system. If the uncertainty reaches some maximum amount of spreading after a finite time, it is not well quantified by global measures derived from Lyapunov exponents (e.g., Lorenz 1965).

For these and other technical reasons, we do not want to use the existence of a positive Lyapunov exponent as THE definition of chaos (though this happens in many textbooks). We want to use such exponents with nuance as indicators of the presence of sensitive dependence of a system on initial conditions.

What positive Lyapunov exponents do tell us is that the dynamics of a system involves the stretching and folding of trajectories, a process that can lead two initially nearby trajectories to diverge away from one another. It is the specific kind of stretching and folding dynamics that picks out a system as chaotic, but this is much harder to mathematically characterize than computing Lyapunov exponents. The stretching of trajectories is associated with the explosive growth in uncertainties while the folding confines all trajectories to a region of state space.

Chaotic dynamics and predictability

Much has been made of chaos raising problems for predictability, particularly in popularizations about chaos. In fact, we have already seen that predictability of chaotic systems is not as bad as feared even if it has its limits.

Return to the idea of iterated maps. Suppose that the initial uncertainty grows by a factor of four on the first iteration, by a factor of three on the second iteration, by a factor 1/3 on the third iteration, by a factor four on the fourth iteration and a factor of two on the fifth iteration. This sounds a bit like the finite time Lyapunov exponents. By the fifth iteration the uncertainty has grown by a factor of 32. The **geometric average** growth in uncertainty over the five iterations is two per iteration. Uncertainty growth is not uniform, but the geometric average gives us some useful information about predictability for our system. Moreover, the growth in uncertainty shrunk on some iterations. Knowing specifics about the dynamics involved in the iterations allows us greater predictability on some iterations, less on others.

For another example, look again at Figure 4.1 for the logistic map. While it is the case that for values of x close to 0 or 1 the uncertainty grows rapidly, for values around .5 uncertainty shrinks (compare the steepness of the rise of the curve close to 0 and 1 with how flattened it is around .5). These growth rates hold true when values of α are such that the logistic map exhibits chaotic behavior.

Another limitation of using Lyapunov exponents for gauging the unpredictability of chaotic systems is when there are two or more variables in state space, such as for x, y, and z in the Lorenz system (Chapter 6). In such cases, the uncertainties in the variables usually are not independent of one another. In other words, the growth (or shrinkage) in uncertainties for the different variables mix with each other, preventing calculating Lyapunov exponents for the growth in uncertainties. Of course, as pointed out above, Lyapunov exponents only tell us about infinitesimal uncertainty growth at best. Finite uncertainty growth is what we see in systems such as Lorenz's. Yet it turns out that for this system we can prove that there are regions in the state space where uncertainties decrease, meaning forecasting the future is quite good in these regions even though the system has regimes that are chaotic (Shen et al. 2018).

In actual-world forecasting cases, scientists deal with finite uncertainties and their growth. They usually do not bother with trying to compute Lyapunov exponents. Rather, they take measurements of the initial state of their system (e.g., Western Europe) yielding ensembles of initial conditions. Using these they can run their forecasting model (they also need a model of the noise in the observations of the initial conditions represented by the ensembles). They then produce an ensemble forecast: On 30% of the model runs using the ensembles, rain is predicted for Hamburg, Germany tomorrow. This goes into the weather forecast on this evening's news of 30% chance for rain in Hamburg tomorrow. Similarly for hurricane forecasting (see Figure 1.3).

The chaotic dynamics of weather systems is a real thing, but we have techniques for coaxing good, useful forecasting for such systems over the near and medium-range future. Your evening weather forecast is still quite useful despite chaos. Similarly for many other types of chaotic systems. The idea that chaotic systems are unpredictable is a myth. That such systems place limits on our forecasting is true (more on this later).

Chapter 6
Strange Attractors

Another feature of chaotic dynamics that has generated much excitement is a different kind of attractor—a **strange attractor**. This is a mathematical object, such as a solution to an equation, having an infinite number of layers of repetitive structure. Imagine taking a magnifying glass to look at any small portion of the attractor and you find that the magnified portion looks identical to the unmagnified region. Now take that magnified region and imaging using another magnifying glass to blow up a small region and you see the identical structure again! Continuous repetition forever of this process would yield the same results. This kind of structure is a

fractal—an object having fractional dimension (more on this in a moment).

Lorenz's 1963 paper is the earliest published example of a strange attractor. The dynamical system he investigated involved three state variables: x representing the speed of a rotating fluid, y representing the temperature difference throughout the fluid, and z representing the deviation of the temperature profile from a linear one. Among many discoveries Lorenz made about this system was the presence of a strange attractor that shares some resemblance with a butterfly (Figure 6.1).

Figure 6.1. The attractor for Lorenz's simplified 1963 dynamical model of the atmosphere.

Strange attractors in state space allow trajectories to remain within a bounded region of the space by folding and intertwining with one another without intersecting or repeating themselves exactly. Moreover, such attractors for dynamical systems have infinite layers of repetitive structure all the way down and never ending. **Prefractals**, on the other hand, only have two or three levels of repetitive self-similar structure before the self-similarity breaks down, though they still have **fractal dimension** (hence, the word fractal is retained in the name).

Fractal dimension sounds bizarre. The dimensionality of our everyday experience is characterized by integers. For instance, a point has dimension zero, a line has dimension one, a square has dimension two, a cube has dimension three, and so forth. We live and move in a three-dimensional world, and we draw two-dimensional figures such as squares, so our experience is of a world of integer dimensions.

Mathematically, we can generalize our intuitions regarding dimensionality, however. Consider a large square. Fill this large square with smaller squares each having an edge length of ε. Counting the number of small squares needed to completely fill the area inside the large square would yield $N(\varepsilon)$. Now repeat this process of filling the large square with small squares, but each time let the length ε get smaller and smaller. In the limit as ε approaches zero, the ratio $\ln N(\varepsilon)/\ln(1/\varepsilon)$ will equal two just as we would expect for a 2-dimensional square. Imagine the same exercise of filling a large 3-dimensional cube (a room, say) with smaller cubes of edge length ε. In the limit of ε approaching zero, you would calculate the dimension to be three.

When we apply this generalization of dimensionality to the geometric structure of strange attractors, we get noninteger results. Roughly this means that if we try to apply the same procedure of "filling" the structure formed by the strange attractor with small squares or cubes, in the limit as ε approaches zero the result is noninteger. Whether examining a set of nonlinear mathematical equations or analyzing the data from a time series generated by an experiment, the presence of self-similarity or noninteger dimension are indications that the behavior of the system under study is dissipative (nonconservative) rather than Hamiltonian (conservative).

A warning is in order here: Just because a fractal is detected in a time series does not mean that the dynamics of the system generating that time series is deterministic. Suppose we take the unit interval from zero to one (imagine a line of all the real numbers from 0 to 1). Now subtract the middle third of this interval to produce a new interval. For the next iteration, remove the middle third of the remaining lines to produce a new interval. Continuing this procedure, as the number of iterations approaches infinity you will construct the **Cantor set**. Following this procedure, you would get a fractal limit of the dimension of approximately 0.631, smaller than what would be expected for the dimension of a line—one. It is possible to produce a nondeterministic version of this procedure. Hence, if one were to look at the fractal produced in this fashion not knowing the procedure, it would be unsafe to conclude that a deterministic process had produced the fractal.

By the way, this same example shows that detecting a fractal in a time series does not guarantee that the system producing it is chaotic. While dissipative chaotic systems will have fractal attractors, not all dissipative systems having fractal attractors will be chaotic.

Chapter 7

Chaos in Conservative Systems

Dissipative systems, those that do not conserve energy, are the norm in physics, chemistry, and biology. As we saw in the previous chapter, the kinds of attractors in these systems have the property that trajectories approaching them contract (either in area or volume) in state space. Conservative systems, by contrast, conserve energy and their attractors are such that trajectories approaching them conserve the area or volume in state space. One thing both kinds of systems have in common, important for chaotic dynamics, is **confinement**: Trajectories near the attractors stay near the attractors.

Relatively speaking, there are far fewer conservative systems in nature, but there are plenty of conservative models. In fact, in chapter 4 we investigated the properties of one of the most well known—the logistic map. Chaos in conservative systems may not have the fractal geometry of strange attractors, but it clearly has intricate behavior. Is there a counterpart in the chaotic dynamics of conservative systems to strange attractors in dissipative systems?

More complex dynamics

The **Chirikov map**, also known as the Chirikov standard map historically, is a more complex system than the logistic map, but is conservative, or state space area preserving. It represents a rotor (think of a stick) that is periodically kicked at one end while the other end is fixed and can rotate free of friction. There is no gravity in the system. The map involves two variables, representing the angular position of the rotor swinging around in a circle, and representing the rotor's momentum:

$$P_{t+1} = P_t + K \sin\theta_t$$
$$\theta_{t+1} = \theta_t + P_{t+1}$$

The parameter, K, measures the magnitude of the kick while $\sin\theta$ is the trigonometric function sine, which takes values of θ from 0 to 2π. Therefore, the map tells us how the momentum and angular position of the rotor at time $t+1$ depend on their values at the previous time step t plus the effect of the kick.

As described, this mathematical model sounds a bit like an unrealistic fantasy. But elementary particles in a circular accelerator such as Fermilab or CERN behave very close to the Chirikov map. The particles are periodically kicked by an electromagnetic field to accelerate them as they move around in a large circle and there is no friction on the particles. Gravity cannot be turned off, but its effects are negligible on protons and electrons. Hence, just as with the logistic map, the Chirikov map is a simplified but reasonable mathematical model to study for gaining insight into a physical situation.

The first thing to notice about the Chirikov map is that it has a fixed-point attractor when $P_{t+1}=P_t$ and $\theta_{t+1}=\theta_t$. These conditions are satisfied when P is zero and θ has the value of either zero or π. These two fixed-point attractors are stable as long as $K=0$, which makes sense because there is no kick to disturb the rotor. If we set the rotor in motion, maintaining $K=0$, we get periodic orbits represented by a dot at those coordinates if we were to plot them. Not terribly interesting for our purposes.

More interesting is to plot a variety of initial conditions for $K=0$ as in Figure 7.1. This figure is known as a Poincaré surface of section plot, where the trajectories are cut by a perpendicular plane so that each dot in the figure represents a trajectory coming out of the page at you. In terms of particles moving in an accelerator, think of the particles as always making a closed loop while you are observing the particles head on as they pass by.

Depending on the initial conditions, the Chirikov map produces either periodic or quasi-periodic orbits when there is no kick. The periodic orbits are closed loops and show up in Figure 7.1 as equally spaced dots. Each of these dots represents a particle trajectory repeating itself exactly. The quasi-periodic orbits are also closed loops and resemble necklaces where the beads are bunched together. For quasi-periodic orbits, the circumference of the loop expands and contracts such that the particles do not repeat their trajectories exactly but still exhibit a regular pattern to their repetition.

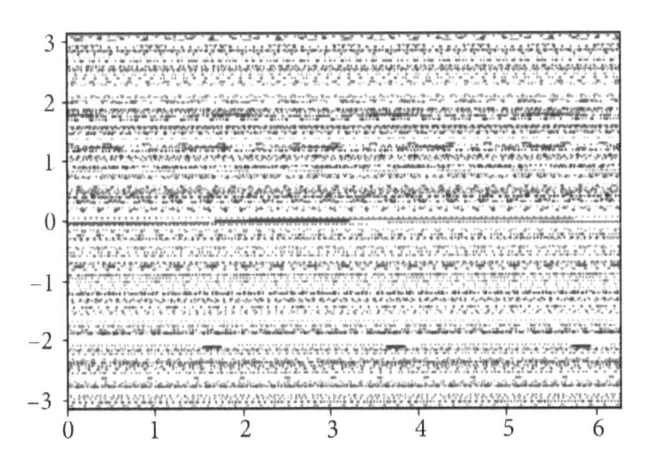

Figure 7.1. The Chirikov map plotted for K=0.0 displaying periodic and quasi-periodic orbits.

Naturally, things change when K is nonzero. It can be shown that the fixed-point attractor P=0, θ=π is stable for values of K ranging from zero to 4. For larger values of K, the fixed-point attractor is unstable. In contrast, the fixed-point attractor P=0, θ=0 is unstable for all values of K greater than zero.

Figure 7.2 illustrates how interesting the Chirikov map behavior is for K=0.5. All periodic orbits look like ellipses, and you can see several of these throughout the diagram. These structures are typically called **islands**, as in islands of stability. If you look closely scanning horizontally across the figure for the value P=2, you can see several small islands of periodic orbits. Meanwhile, the quasi-periodic orbits now look like necklaces where the beads are wavy. These are all effects from the nonlinearity (Chapter 2) due to the term in the model involving K. Notice that where P=0 and θ=π (about 3.142) you see an empty space that resembles a fat x. This is the unstable fixed-point attractor, and all trajectories move away from this point.

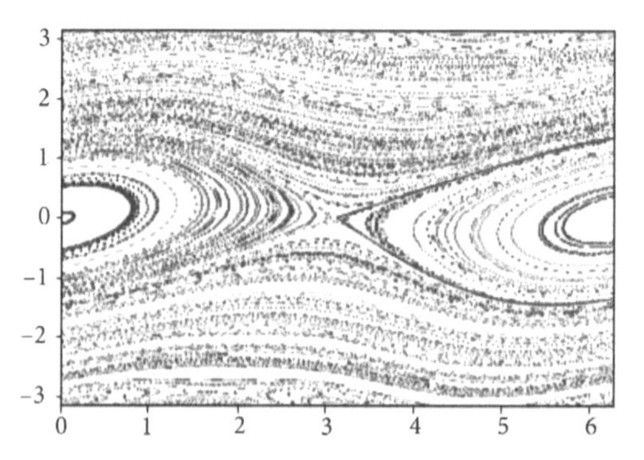

Figure 7.2. The Chirikov map plotted for *K*=0.5 displaying periodic and quasi-periodic orbits

Increasing K to 0.971635, Figure 7.3 reveals that while there are several islands of periodic orbits, there are very few quasi-periodic orbits. In addition, there are several dots that appear to have no pattern. These are the aperiodic orbits that are the signature of chaotic behavior. A slight change in initial conditions, given a value of K above a critical value, leads a quasi-periodic trajectory to become aperiodic.

Mathematically, these aperiodic orbits represent trajectories that never repeat themselves (so far as we can tell). This nonrepeating behavior is the reason why the points representing aperiodic trajectories appear to have no discernible pattern—they are wandering around filling state space. Nonetheless, since this is a conservative system, no trajectories ever cross each other; even the chaotic trajectories never cross themselves or any others. The critical value of K at which chaotic trajectories first appear is not known, but it is thought to be close to 0.97.

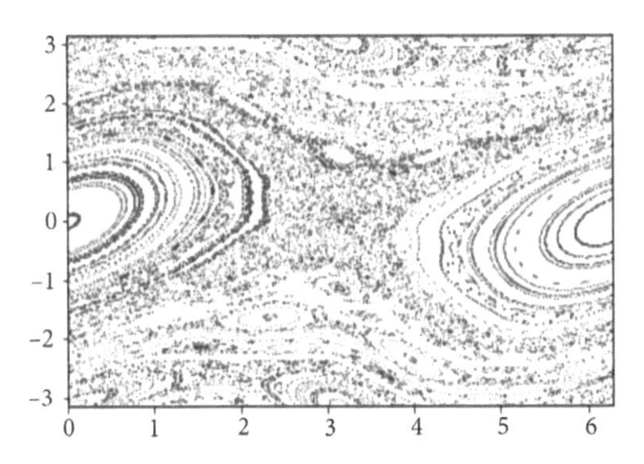

Figure 7.3. The Chirikov map plotted for K=0.971635, where some aperiodic orbits appear among the periodic and quasi-periodic orbits.

Figure 7.4 shows what the behavior looks like when K=5.0. The figure looks like a sea of chaotic trajectories surrounding islands of stability. One can say the same thing for Figure 7.3 as well. This is a general feature of conservative maps exhibiting chaotic behavior.

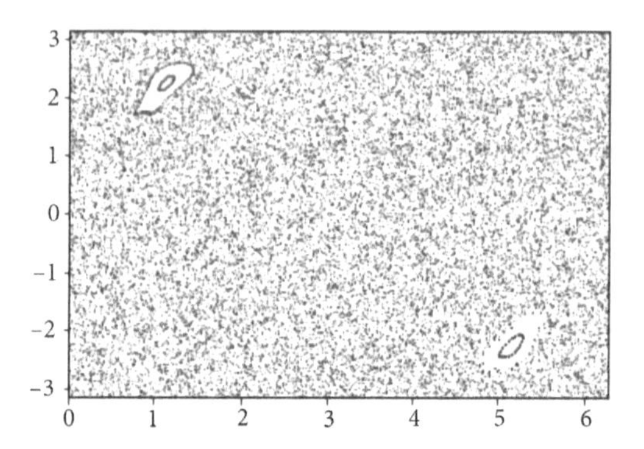

Figure 7.4. The Chirikov map plotted for K=5.0. There are always islands of stable periodic trajectories among the chaotic, aperiodic trajectories.

Notice that the fixed-point attractor at P=0, θ=0 is no longer stable. Recall this attractor is stable for values of K less than 4. At K=4, the attractor bifurcates into a periodic loop attractor with period 2. As K increases, this period-2 attractor will undergo another doubling bifurcating into a period-4 attractor and so forth when technical conditions are satisfied that I cannot go into here. This process of periodic loop bifurcation continues, eventually producing a period-3 attractor and so forth leading to a sequence of periodic loop attractors of periods 1, 2, 3,... yielding a structure of islands of stability surrounding islands of stability. These self-similar structures occur for all periodic loop attractors satisfying the technical conditions in the previous figures. This is the counterpart to the behavior of strange attractors with their repeating structures on smaller and smaller scales for dissipative maps. Sometimes fractal structure can appear in conservative as well as dissipative models. It is just that the fractal structures in conservative models are not associated with strange attractors. This may surprise you, but the sequence of period doubling is described by the same Feigenbaum constant as with the logistic map.

Having examined the Chirikov map in some detail, look again at Figures 4.8 and 4.9 for the logistic map and compare them with what you have seen for the Chirikov map. Just as with the latter, the former has many stable periodic orbits in the midst of a sea of aperiodic orbits. Again, this is just what we expect for a conservative or area-preserving nonlinear map. Furthermore, there is an infinite number of these periodic orbits. Just as with the logistic map, the Chirikov map has periodic attractors of periods 1, 2, 3,... There is an infinite number of them because there are infinitely many rational numbers corresponding to orbits of period *n*. And as with the logistic map and the Lorenz system, there are far more stable periodic than aperiodic orbits in the Chirikov map.

Taking stock

Stepping back, we can draw some observations from the previous chapters. First, claims that chaos is ubiquitous are misleading. Nonlinearity is ubiquitous, but as you have seen, chaotic dynamics is not always present in nonlinear mathematical models. As both the logistic and Chirikov maps illustrate, chaotic dynamics only occurs when the conditions are right for it.

Second, the aperiodic orbits in the logistic and Chirikov maps, and in the Lorenz system exhibit apparent randomness (see the Introduction). Remember that the mathematical equations for all three model systems are deterministic. It turns out that deterministic dynamical systems are capable of exquisite and surprising order.

Third, as previously mentioned, this all has implications for predictability of chaotic models. Namely, they are far more predictable than the popular rumors have it. For one thing, nonlinear models exhibit nonchaotic behavior for far more parameter settings than chaotic. Yet, even when there is chaotic dynamics, we can predict a great deal of the intricately ordered behavior of these systems. The real issue with the sensitive dependence on initial conditions associated with chaotic dynamics is handling the growth in uncertainties, as you will see.

Chapter 8

Physical Systems and Chaos

So far, the focus primarily has been on chaotic behavior of mathematical models. What about chaos in the physical world? Fascinating as the chaotic dynamics of mathematical models, such as the Chirikov map, is, it is actual-world behavior we ultimately want to understand. Much of the discussion in this book has focused on relatively simple, low-dimensional systems such as the logistic map or Lorenz's early three-dimensional model. Do such low-dimensional mathematical models capture realistic features of the physical world? After all, it could be the case that our mathematical models exhibit all kinds of curious behavior that are never seen in the actual world.

I already introduced you to an example where our mathematical models exhibit curious behavior not found in the actual world: fractals. Strange attractors in our mathematical models have fractal nature, as does the self-similar series of periodic attractors in the Chirikov map. Yet, we also find some self-similar repeating structure in the physical world. Consider the Atlantic coastline of America. Now zoom in to focus on the coastline of North Carolina. Statistically, the latter coastline is self-similar to the American Atlantic coastline. Zoom in again to Coquina Beach. Statistically, the beach shoreline is self-similar to North Carolina's coastline. Though, in contrast to mathematical models, physical self-similarities are always prefractals with statistically self-similar structure usually only repeating on two to three spatial levels.

The fractals in our mathematical models live in the state space of the model while physical prefractals live in the actual world. Here is a clue that our mathematical models can have far more intricate structure than is found in the physical world. In the case of fractals versus prefractals, the former have infinite self-similar behavior in our mathematical models while the latter typically only exhibit forms of self-similar behavior on two to three spatial levels.

Or consider Lyapunov exponents. It is relatively straightforward to calculate these exponents for simple mathematical systems, such as the Logistic map. It is more difficult to measure them for dynamical systems such as Lorenz's equations. And it is harder still—perhaps even impossible sometimes—to measure Lyapunov exponents for physical systems.

These examples hint that there may be some questions about the faithfulness of our mathematical models to the actual world. We will return to this possibility in a moment.

Nevertheless, here is something astounding. Recall the Feigenbaum constant computed for the values of the parameter α for successive bifurcations in the Logistic map in Chapter 4: -4.6992016091. It turns out that when scientists compute the Feigenbaum constant for period doubling in physical systems such as water, mercury, and liquid helium systems, or in chemical reactions such as the visually stunning Belousov-Zhabotinsky reaction, or in lasers, diodes, and transistors, these computed values are very close to the value for the Logistic map. Yet this map does not describe the behavior of these quite different physical systems.

Numbers, computers, and the physical world

One nuance in the relationship between mathematics and the physical world is found in numbers. Recall there are two kinds of numbers: integers and reals. An **Integer number** is a whole number such as the winning score in a football match. A **real number** can have an infinitely long string of digits to the right of the decimal place such as the value of π which is 3.14159…, the dots indicating the rest of the infinite string.

Our mathematical models involve both integer and real numbers. Physical measurements, on the other hand, always involve integers. "But hold on," you might say, "Don't we usually measure height, weight, and other quantities with digits to the right of the decimal place?" Fair question. For example, my height measures to 1.778 meters. As with this measurement, we never measure out to an infinite number of decimal places. My measured height is an integer in disguise. Just multiply by 1000 and you have an integer: 1778. Scientists say that all measurements have **finite precision**, whether they be height, temperature, atmospheric pressure, mass or whatever. Hence, all our measurements are really integers in disguise.

Not only this, but all the numbers in our computers are represented as integers as well. All distances, speeds, temperatures, and other quantities measured by any digital devices are all integers and stored as integers in hard drives and other memory devices. The world of mathematical models is a world of integer and real numbers, while the digital world is a world of integers only.

Furthermore, when scientists make measurements, they are always of the form 1.778 ± 0.005 meters, say. This quantifies the uncertainty in the measurement of my height. This uncertainty is due to jitter in the apparatus used to measure my height as well as my hair and scalp and the bottom of my feet, and so forth. My spelling out some of the sources of jitter—a form of what scientists call noise—indicates that scientists are always working with a model for the noise (its sources, structure, etc.). A very common model to use for noise as at least a first approximation is a Gaussian or bell curve (also known as a normal curve) as in the example in Figure 8.1.

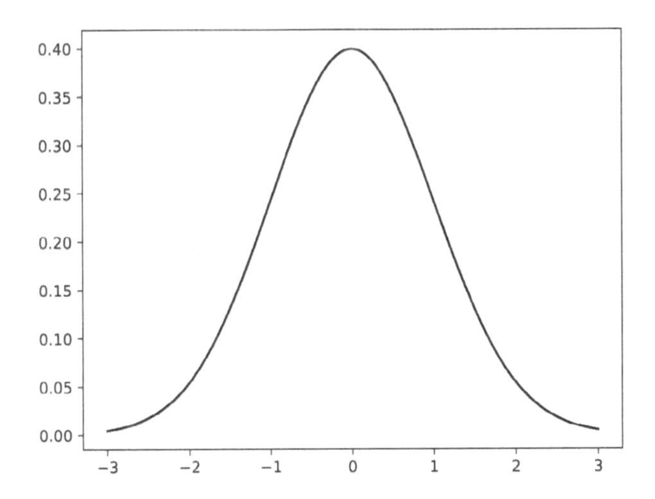

Figure 8.1. A Gaussian or bell curve. The area under the curve is equally distributed around the peak.

A bell curve represents a distribution of values (noise) around the center of the peak. Scientists often use such a model for noise when they have no indications that uncertainty is distributed unevenly or asymmetrically. For example, the error bars—

the ±—in my height measurement indicate that the uncertainty is distributed evenly around the central value, 1.778 meters.

In the world of mathematical models, however, there is no uncertainty—no noise. This is the world of perfect models and exact precision to infinite decimal places. For instance, when we set $\theta = \pi$ in the Chirikov model, the symbol π includes the infinite number of decimal places. The world of mathematical models is a very different world than that of digital computers. And perhaps also quite different from the actual world's physical properties such as my height or the temperature at Zürich Airport. Is there a true value to such magnitudes as my height or the temperature at Zürich Airport, one might ask? I will call these True values, values that exist independent of any measurement and noise. Scientists have learned a lot about uncertainty and noise and whether True values exist in our world by studying chaotic dynamics, as you will see.

Generally, scientists define uncertainty as two types. First is uncertainty as a range of values produced by a measurement. Take the example of my height measured as 1.778 ± 0.005 meters. In scientific contexts, this kind of observational noise comes from the fact that no measuring instruments are perfectly accurate.

Second is uncertainty in our computer or model representations. If you input 1.776 meters into the computer to represent my height but my True height is 1.778 meters, then there is an error in the representation—your statement—of my height in the computer. In scientific contexts, representational uncertainty is more complex. Suppose we want to accurately estimate the average height of football players in Africa. It is not possible to measure the height of every player, so we sample the heights of several players from different countries and average those to get an estimate. The more players we measure, the better our average estimate. But this average estimate will always be an imperfect representation. The representation contains some error. This means our models for uncertainty will always contain some inaccuracy.

Returning to the weather forecasting example, typically scientists measure quantities such as temperature, pressure, and humidity in some region as data to input into their forecasting models. These measurements lack perfect accuracy to begin with. Furthermore, measurements always return integer values rather than real numbers (we believe that quantities such as temperature and pressure are real numbers in the physical world). This introduces additional uncertainty into the initial conditions. **Data assimilation** is the process of mapping or transforming the observations into the model state space. In essence, we now have a model of the observations—the initial data—within the model state space for our forecasting model to use as initial conditions. The assimilation process introduces some further uncertainty into the initial data. Scientists represent the total uncertainty in a model for the error between the measurements, as represented in the model state space, and what the True values of these quantities are believed to be in the actual world. This model for data uncertainty needs to be both consistent with the observations and consistent with the weather forecasting model if scientists are to make a reliable forecast for people to use.

Models, faithfulness, and the actual world

A further word about models and their subtleties. First, the word 'model' may bring to mind toy models from childhood, or people who promenade down a catwalk or pose for photographs. In this book, I have been using the term in the way scientists mean: a simplified mathematical description of the key variables and processes of a system under study. Yet, as toy-like or artificial as models may sound to you, they can have significant actual-world consequences. For example, models are constantly used in economics and public policy, public health (e.g., in the Covid-19 pandemic), weather forecasting, and in anthropogenic global warming leading to climate change, among other contexts. Policy makers are using such models to make decisions that affect our daily lives as well as lives of future generations. Therefore, it is important that we understand what we are doing when we are using models to understand how nature works and as guides to our decision making.

There is a tight connection between the model used to produce a weather or Sunspot forecast and the state space of variables thought important to track (Chapter 3). However, note that there is an assumption being made here that is often passed over: Scientists assume the state of the target system of interest—a weather system or the Sun, say—is appropriately characterized by the values of the state space variables. This implies that the system state represented in state space corresponds directly to the state of an actual-world system through these values.

This assumption allows us to develop mathematical models for the evolution of these points in state space and such models are taken to faithfully represent the target system. Call this the **faithful model assumption**. The equations are taken to faithfully represent important features of the target system, while the state space is taken to faithfully represent the target system's actual space of possibilities. Given this faithfulness assumption, one can reasonably think of plotting trajectories for both the target system and its model in the same state space. Nevertheless, I have already given you some reason to believe that this faithfulness is delicate. Since real numbers exist in the actual world and the mathematical model world, the state space must have the same reals. But when we plot the values of our actual-world system in the state space, we can only plot the integers we have measured with finite precision. This raises questions about how faithfully we can represent the actual-world target system in the state space to compare with the mathematical model.

The most idealized form of the faithful model assumption is where we have perfect mathematical models. Our models are taken to be perfect representations of target systems or, as is more often the case, such perfect models are assumed to exist and analysis is carried out within this perfect model framework even though our models may be flawed in several ways.

One advantage of the perfect model framework is that it licenses (often sloppy) switching back and forth between model talk and target system talk. After all, if a climate model is taken to be a perfect representation of the Earth's climate, then any statements made about the model should also be true of Earth's climate. Nevertheless, using the perfect model framework when we do not have perfect models is problematic, particularly in contexts where we are dealing with nonlinear effects.

The reality is that scientists have neither perfect models nor perfect data. What scientists generally end up doing is following piecemeal strategies for improving our models and data for actual-world systems (representing competing approaches vying for government funding; for an early discussion in weather forecasting, see Thompson 1957).

The first basic approach is to focus on piecemeal improvements to the accuracy of the data fed into the model while keeping the model fixed. The idea here is that if a model is faithful in reproducing the behavior of the target system well, improving the precision of the data fed to the model will lead to its output monotonically converging to the target system's behavior. This is to say that as the uncertainty in the initial data is reduced, a faithful model's output is expected to converge to the target system's behavior. This is the invocation of the faithful model assumption allowing one to plot the trajectory of the target system in an appropriate state space. We would expect to see the target system trajectory and the model become monotonically more alike as data accuracy is improved.

The second basic approach is to focus on piecemeal improvements of the model while keeping the data fed into the model fixed. The idea here is that if a model is faithful in reproducing the behavior of the target system to some degree, refining the model will produce an even better fit with the target system's behavior. Invoking the faithful model assumption, if one were to plot the trajectory of the target system in an appropriate state space, the model trajectory in the same state space would monotonically become more like the system trajectory as the model is made more realistic.

What both basic approaches have in common is piecemeal monotonic convergence of model output to target system behavior under the faithful model assumption. For linear models, it is easy to see the intuitive appeal of such piecemeal strategies. After all, for such models a small change in the magnitude of a variable is guaranteed to yield a proportionately small change in the output of the model. By making piecemeal improvements to the initial data or to the linear model, only proportional changes in model output are expected. If the linear model is faithful, then making small improvements "in the right direction" in either data or model accuracy can be tracked by improved model performance. The qualifier "in the right direction," drawing upon the faithful model assumption, means that the data quality really is increased or that the model really is more realistic (captures more features of the target system in an increasingly accurate way), and is signified by the model's monotonically improved performance with respect to the target system comparing the two trajectories in state space.

In contrast, for nonlinear models where sensitive dependence can become an issue, these piecemeal approaches are problematic when we have imperfect models but treat them in the perfect model framework. Suppose we improve the initial data fed into nonlinear models to some degree. Such improvement is not guaranteed to lead to more convergence between imperfect model behavior and target system behavior. Due to sensitive dependence, any small refinements in the quality of the data "in the right direction" are not guaranteed to lead to the nonlinear model monotonically improving in describing the target system's behavior. The small refinement in data quality may very well lead to the model behavior diverging away from the target system's behavior.

In the second approach, keeping the data fixed but making successive refinements in nonlinear models is also not guaranteed to lead to more convergence between model output and target system behavior. Because of sensitive dependence, any small changes in the model can lead to non-proportional changes in model behavior, even if made "in the right direction." There is no guarantee that the nonlinear model will monotonically improve in describing the target system's behavior. The small refinement in the model may very well lead to the model behavior diverging away from the target system's behavior (I have witnessed this in plasma physics modeling).

There is a further issue with the faithful model assumption itself, namely what is the relationship between the model and the actual-world target system? Scientists typically assume this relationship is one-to-one. This means we can map parameters and variables and how they change with time in the model equations to actual features of the target system and vice versa. But it may be the case that the actual relationship is one-to-many. In other words, several different nonlinear models of the same target system or, potentially, vice versa. Or maybe the mapping between models and actual-world systems is a many-to-many relationship. For linear systems such as a pendulum where the bob is executing small swings, the mapping or translation between model and target system appears to be straightforwardly one-to-one. However, in nonlinear contexts, particularly where one might be constructing a model from a time series generated by observing a system, there are potentially many nonlinear models that can be constructed. Each model is as empirically adequate to the system behavior as any other. Is there only one unique model for each target system and we simply do not know which is the True one? Or is there really no one-to-one relationship between our mathematical models and target systems? Scientists and philosophers of science do not know the answer to these questions. The work of the scientist seeking to model actual-world phenomena is hard and challenging indeed.

As a last comment on modeling, quite often one reads scientists talking about how equations, models, and laws "govern" the target systems' behaviors. This is the **governance myth**. Equations, models, and laws do not govern or control anything. Consider an analogy: The number π characterizes the geometry of disks and balls but it does not control what disks and balls are or how they behave. It does represent a constraint on their shapes. Similarly, the universality of the value of the Feigenbaum constant across so many mathematical one-hump maps and disparate physical systems does not imply that it possesses any governing control over these systems. What this universality tells us is that a wide variety of different systems somehow share some key structural features.

We can never prove that our equations, models, or laws govern anything, but only demonstrate that they describe actual-world phenomena well (e.g., capture the constraints on a phenomenon's behavior accurately such as Newton's law of gravity describing the acceleration of a falling brick).

Sensitive dependence and measurement

There is another implication of sensitive dependence to consider for measurement raised by nonlinear models relevant to thinking about chaos in physical systems.

As I have emphasized throughout this book, no measurements are perfect. Every measurement has error bars attached to it such as my height measured to be 1.778 ± 0.005 meters. A typical assumption, particularly under the perfect model framework, is that it is possible in principle to reduce any such errors to be as small as is needed so any measurement can ultimately be as accurate as needed. But this is more a hope than an argument. In fact, scientists have no justification for helping ourselves to the assumption that any measurement errors can be arbitrarily reduced.

Suppose for the sake of argument that someone could devise a measurement process for physical systems that can be made arbitrarily accurate. Is perfect measurement accuracy attainable with such a process? No, because error reduction is a limiting process. As the measurement process is refined, at some point the information storage needs will exceed the storage capacity of all the universe's storage devices. Because of sensitive dependence, to have perfect measurement accuracy requires infinite precision. But since the information needed to account for accuracy grows exponentially, the storage capacity of the universe will be exceeded long before a measurement can reach infinite accuracy.

Moreover, there is an additional source of uncertainty in our measurements that often cannot be mitigated. Measurements with instruments almost always introduce some small disturbance into the system being measured. Hence, even if someone could continually refine measurement error as small as desired, the act of measurement with devices itself can disturb the system being observed. And even the smallest disturbance introduced through the measurement introduces some uncertainty into the True state of the system. Sensitive dependence implies this uncertainty will grow rapidly with time.

But it is not just the act of measurement scientists need worry about. Recall the idea of the flap of a butterfly's wings disturbing air molecules in Brazil leading to a tornado in Texas three weeks later. There are lots of possible small disturbances that a physical system exhibiting sensitive dependence could amplify. For example, it is possible for an electron at the greatest distance known in the universe to affect a system of billiard balls undergoing continuous collisions exhibiting sensitive dependence (Crutchfield 1994). In principle, a system exhibiting sensitive dependence could amplify the uncertainty introduced by small disturbances due to the minute fluctuations in gravity and electromagnetic fields owing to the movement of scientists and their vehicles. To make a complete accounting for all these effects on the chaotic system adds to the needs for storage capacity in the universe to mitigate all the possible sources of disturbances, adding to the uncertainty in the initial conditions for the chaotic system scientists want to study.

The upshot of all of this is twofold. First, the perfect model framework underestimates how difficult it is to have anything like a perfect model for the uncertainty when modeling nonlinear systems that are exhibiting chaos. Therefore, scientists generally exercise care when analyzing and thinking about modeling in actual-world situations where such perfect models do not exist. Faithful models are the best we can hope for, though we are still left with our questions about how faithful our nonlinear models are to actual-world systems.

Second, there are further constraints on how long one can predict the future behavior of chaotic physical systems. This makes it even more important to be able to discern when systems are in a chaotic regime and how to extract useful information from our forecasting of such systems (to be further explored below).

Chaos in the actual world

Recall the example of the damped driven pendulum in Chapter 2. From our mathematical model we can show that this mechanical system exhibits chaos for some combinations of the magnitude of friction and magnitude and frequency of driving force. Suppose we build an actual pendulum system and set all the parameters according to our model predictions. What do we find?

Initially, we find that the pendulum behaves chaotically as the model predicts. However, in the actual world, the friction at the pivot point heats up the materials changing the magnitude of the working coefficient of friction. Chaotic behavior can disappear because of the change in friction. The magnitude of the coefficient of friction in the mathematical model is a constant whereas in the actual material system the magnitude changes with time (due to several microscopic processes). Similarly, the motor used to drive the pendulum heats up over time and the magnitude and frequency of its driving action can begin to vary with increasing temperature. These parameters are treated as constant in our mathematical model.

The bottom line is that chaotic behavior is persistent in our mathematical model while it is relatively short lived in the physical pendulum system. This illustrates how our mathematical models can often be inadequate representations of the material world even when the relevant variables and parameters are included.

Earlier, I discussed the difficulties defining chaotic dynamics as the presence of a positive Lyapunov exponent in a system's time series (not the least of which is the difficulty of calculating these exponents for time series data from actual-world systems such as our mechanical pendulum). The suggestion in Chapter 5 was to pay attention to the stretching and folding dynamics of systems to look for sensitive dependence on initial conditions (and parameter values). While scientists and philosophers may lack rigorous mathematical definitions for stretching and folding dynamics, scientists and engineers do have ways of investigating physical systems to understand the extent to which such dynamics are taking place.

Indeed, stretching and folding dynamics, along with determinism, are necessary conditions actual-world systems must possess if they are to be capable of behaving chaotically. Nevertheless, these conditions are not sufficient to guarantee systems will always exhibit chaotic dynamics.

Now suppose that actual-world systems such as the weather possess the necessary and sufficient conditions that qualify as chaotic. What does that mean in practice? As the hurricane example in Chapter 1 illustrates, it means that any imprecision in the measurement of an earlier state of a weather system will lead to much larger uncertainty in what the forecasted weather will be like in future states.

This is an **epistemic issue**—a problem of our current versus projected future knowledge of the weather. Along with irremovable uncertainty from our measurements, it is also important to emphasize that whatever the actual earlier state of the weather system is, the system will evolve based on that state, not on our knowledge about that state. What is important to the physical behavior of the weather is the nature of the attractor in the stretching and folding dynamics of its fluid dynamics, not how precisely we can know its earlier state. Hence, the apparent randomness we observe in actual-world systems (see the Introduction).

The practical implication of the weather or other actual-world systems being chaotic is the limitations on our ability to predict the future, as illustrated several different ways in this book. Nonetheless, scientists can extract useful forecasts from nonlinear systems such as the weather if we are alert to the lessons chaos is teaching us.

Chapter 9

Using Computers Wisely

Computers are central to the work of mathematicians and scientists. We run models such as Lorenz's on them. We store and analyze time series data with them. We build models of such data using computers. Chaos has taught us how to better use computers to model the physical world.

Computers

The fact that so much modeling work involves computers leads to further nuances relevant to studying chaos. As pointed out earlier, while both real and integer numbers exist in our mathematical models, digital computers can only use integers. This means that there is no way for a computer to represent the difference between numbers if that difference is too small. For instance, if infinitesimals are crucial to chaotic dynamics, computers will treat all such infinitesimally close real numbers as if they are the same integer. This spells trouble for using computers to study chaos. Yet, Lorenz successfully calculated aperiodic solutions that exhibited sensitive dependence. What gives?

Recall that when Lorenz restarted his calculation using computer output from an earlier point in the computation there were truncation errors, meaning that he was feeding slightly different numbers into the computer to represent that earlier state. The computer dutifully (deterministically) computed a different trajectory from this new starting point than it had on the original run.

There is another nuance about digital computers here. Since all computers have a finite amount of memory no matter how large, there is a limit on how many distinct internal states they can have. This means that at some point in any calculation a computer will have to return to a previously used state. As deterministic devices, the computer will begin to repeat the same calculation again. Hence, there is no way with a computer to determine if aperiodic solutions will remain aperiodic indefinitely. Perhaps aperiodic solutions repeat in some limit cycle (Chapter 4), and we have not run our super computers long enough with enough memory to find this cycle yet.

On the one hand, for most of the complex problems scientists study, computers are an unavoidable tool. We can see that the scientist's task of modeling actual-world systems is very challenging. On the other hand, scientists use computers, as Lorenz did, to study and gain understanding about nonlinear physical systems with great success. Here, success means that the output of our computer model **shadowed** the time series generated by observing the target system closely enough to be consistent with the model for noise for a sufficiently long enough time. And remember that the noise model includes both the uncertainty in the data fed into the computer model as well as the representational error introduced in the assimilation process (Chapter 8).

One of the lessons chaos has taught us is that scientists must pay closer attention to our models of noise than we were used to doing when studying linear systems. Given that models of target systems are imperfect and models for noise are imperfect, we know that there are limits to how long the **shadowing time** will be. That is, how long our computer model output will remain reasonably close to the target system trajectory in state space.

There is always some uncertainty/inaccuracy in the initial data fed into our computer models, and we are limited to integer representations. Moreover, our measurements typically contain information on smaller scales than we can model in even the most powerful supercomputers, so this gets left out of the initial data fed into the computer model as well as the model itself. Not to mention that there are variations in the key variables for weather on even smaller scales than our measurements can detect. There are always temperature and pressure differences and eddies in wind currents on length scales that are too small to ever get into the computer model. Furthermore, there are the questions about the relationship between our models and actual-world systems from the previous chapter. If we are dealing with chaotic dynamics, then we know there are nonlinearities in some form. And when we have nonlinearities, we need appropriate techniques for analyzing models and data because our techniques for linear systems usually fail in the context of nonlinearity. This is where shadowing and ensemble forecasting enters.

The Shadow knows?

Chaotic dynamics does not render predictability hopeless though it does place some limits on our ambitions. Producing probability forecasts is an appropriate ambition. The forecast for hurricane Elsa in Figure 1.3 is an example of a probability forecast. The probabilities produced by the model reflect the growth in uncertainty from the time the latest observational data is assimilated into the model to the forecasted times and possible locations of the hurricane's landfall. A key assumption is that somewhere in the model's evolving cone of uncertainty lies the actual trajectory the hurricane will travel. In other words, the True state of the hurricane is located somewhere in that probability distribution.

This is the idea of shadowing. Out of the ensemble of initial conditions within the uncertainty of the observations forming the initial data assimilated into the computer model, there should be at least one initial state under iteration such that the model's computed future states stay close to those of future observations. In other words, the model's time series output tracks closely with or shadows the time series of the observations NOAA, say, is taking of the evolving storm. A model adequate to our forecasting needs will have at least one such shadowing state, and this state and its evolution will be reflected in the ensemble forecast.

This suggests a forecasting strategy: (1) Collect the initial observations of the storm system and assimilate these into the forecasting model. (2) Make several model runs producing an ensemble forecast, varying the initial data for each run within the uncertainty of the observations. (3) Compare the probability forecast with the next set of observations of the storm. (4) Keep those model states that shadow the observed behavior of the storm. (5) Assimilate the latest observations into the model. Repeat (2)-(5). Following such a strategy keeps our model accountable to all the data up to the present while giving us a reasonable projection into the future of the likely times and locations of the hurricane's landfall.

To make such a strategy work with a sufficiently high level of forecasting skill, we not only need a model that tells us how to estimate the minimum size of the ensemble needed to be confident that we have a sufficient number of shadowing trajectories but also techniques for minimizing or reducing the noise in our observations as much as possible. Good (but not perfect) models, statistics, and practices enable NOAA, the European Center for Midrange Weather Forecasting, and other forecasting agencies to produce quite useful probability forecasts that shadow actual hurricanes and other weather systems well enough to save many lives and avert some economic damage.

Modern weather modeling (as well as modeling for climate and other physical systems) also makes use of stochastic or random noise. Our computers are not (and likely never will be) powerful enough to run models that incorporate information in the data down to the smallest scales (inches or millimeters, say; quantum mechanics will place a limit on how small the length scales can be that our models can track in principle—see the next chapter). To overcome this limitation, scientists inject some random noise into each model run to account for the information the model cannot track. This is apparent randomness since computer random number generators depend on deterministic algorithms. In essence, we treat the unknowns at these smaller length scales as a kind of random variable in our model. Hence, our computer model for weather also includes a model for the effects we are unable to represent with the computer.

Each model run building the ensemble gets a slightly different noise injection. This means that we are using a model for the noise representing the missing information. Different stochastic injections amount to letting the information missing from our model behave slightly differently on each model run. Such techniques make our ensemble weather forecasts much more accurate than if we left noise out (Palmer 2019). Chaos has taught us that noise can be our friend rather than our enemy.

The latest addition to the forecaster's toolkit is **machine learning**. Machine learning, sometimes called artificial intelligence, is a branch of computer science that develops computer algorithms to solve problems, detect patterns in data, represent and collate knowledge, and so forth. In brief, a machine learning algorithm is designed to take a function that represents the solution to a problem or a pattern recognition task, say, facial recognition. Scientists then use large data sets to train the function to solve the problem (e.g., describe fluid flow through a system of pipes) or recognize the pattern (e.g., when people are crossing the street in front of a moving vehicle).

A machine learning algorithm trained to recognize chaotic behavior can be coupled with weather observations to recognize when a weather system is in a chaotic regime where nonlinearities are having significant effect (Barbosa and Gauthier 2022). Such machine learning algorithms can make their own forecasts based on these observations or can be used in conjunction with ensemble forecasting models so weather forecasters can take chaotic dynamics into account in the construction of their ensemble forecasts.

Chapter 10

Quantum Chaos?

Quantum mechanics is the science of the very small—elementary particles and atoms made up of such constituents. The quantum world is very strange compared to our ordinary experience and intuitions. As such, it has been the subject of countless science popularizations and makes regular appearances in science fiction (though sometimes the science fiction is too weird even for quantum mechanics!).

This book has been exploring chaotic dynamics for the macroscopic world, the world of our experience. What about quantum chaos? That certainly gets talked about a lot and sounds exciting. This may feel like a letdown, but chaos as I have been describing it for the macroscopic realm does not appear to exist at the microscopic, the quantum realm. What scientists typically study is more aptly called **quantum chaology**, the relationship between chaos in the mathematical models of the macroscopic realm and the models of the quantum realm.

Why is chaos as we have been studying it absent from the quantum realm? One reason has to do with determinism, one of the necessary conditions for chaos in macroscopic systems. Quantum mechanics is thought to be the quintessential example of an indeterministic theory. Although there are questions one can raise about the status of indeterminism in quantum mechanics (e.g., Bishop 2006), currently we have no good reasons for doubting that most if not all quantum systems are indeterministic. As such, we do not expect the chaotic dynamics of the previous chapters to exist in quantum systems.

A second reason traces back to our difficulties defining chaos for the macroscopic world. Recall that both our mathematical and physical models exhibiting chaotic dynamics all involved the presence of some form of stretching and folding mechanism associated with a nonlinearity in the system. As this nonlinearity becomes significant, chaos can appear. In contrast, quantum systems are described by Schrödinger's equation and this equation is linear, meaning quantum mechanics is a **linear theory**. This equation implies that quantum states starting out initially close remain close to each other in state space throughout their evolution. There is no exponential separation between nearby quantum states as quantum systems evolve in time. The quantum realm lacks a key ingredient for macroscopic chaos.

Quantum chaology

But this is not the end of the story. What scientists and mathematicians do is study the **quantization** of macroscopic chaotic systems. What this means is they take a nonlinear macroscopic map, such as the Chirikov map, replace its macroscopic variables with quantum ones, and study how these quantized systems behave compared with their macroscopic counterparts. Such systems are often called **semi-classical** as they stand a bit between the quantum realm and the classical realm of the macroscopic. As it turns out, there are many remarkable behaviors exhibited by such quantized systems. It is these behaviors that raise questions about what form chaotic dynamics might take (if any) in the quantum domain and about the relationship between the quantum and macroscopic realms. By the way, the Chirikov map was one of if not the first map to have its quantized version investigated.

Studies in quantum chaology typically focus on universal statistical properties that are independent of the quantum systems under investigation. An important feature of quantum systems, such as electrons in an atom, is their energy levels, so researchers study the statistical properties of the energy levels of systems such as the quantized Chirikov map. The states of quantum systems are described by **wave functions**, hence researchers are interested in the structure of wave functions for these quantized systems. These statistical properties are relevant for quantum state transitions, ionization, and other quantum phenomena found in atomic and nuclear physics, solid state physics that makes your cell phones possible, and even quantum information.

Billiards are a particularly well-studied family of models in this context. Think of a perfectly flat billiard table and assume that the billiard balls bounce off the edges of the table without losing any energy or speed. Such a model table at the macroscopic scale of our experience, where the balls and edges are characterized by Newtonian physics, is called a classical billiard. As mentioned in Chapter 8, billiards systems can display sensitive dependence on initial conditions and have been extensively studied as examples of chaotic dynamical systems.

A chaotic billiard is a classical billiard where the conditions lead to chaotic behavior of the balls. Given the wealth of results for chaotic billiards, quantum versions of billiards have become workhorses for studying quantum chaology as discussed below. One can produce quantum billiards by using Schrödinger's equation to describe the wave functions of particles reflecting off the boundaries (where one specifies that the wave function for the particles is zero at a boundary), or one can start with the equations describing a classical billiard and quantize the observables (e.g., position and momentum) yielding quantized billiards.

Isolated quantum systems

Here are two more relevant differences between macroscopic systems exhibiting chaotic dynamics and quantum systems. First, the state spaces of macroscopic systems support fractal structure while quantum state spaces do not.

Second, macroscopic chaotic dynamical systems have a continuous energy spectrum associated with their motion; in other words, the system's energy can take on any value, such as that of a real number. It makes no difference whether macroscopic systems are bounded (e.g., planets orbiting the Sun) or unbounded (e.g., the Voyager probes leaving the solar system). Nevertheless, as noted earlier, one of the necessary conditions for chaos is that systems be bounded. In contrast, while unbounded quantum systems can have a continuous energy spectrum, bounded, isolated quantum systems have a discrete energy spectrum associated with their dynamics; in other words, the system's energy can only take on integer multiples.

So, what do scientists observe when these classical maps exhibiting chaos are quantized? The latter do exhibit interesting statistical properties, but none of these properties have the signatures of chaos. For example, particles exhibiting chaotic dynamics in classical billiards models have correlations, while the wave functions in the quantized analogs of these models have no correlations. The quantum wave functions are random, meaning their statistical properties differ from the particles of their macroscopic counterparts.

In the case of the Chirikov and other maps, when these are quantized, the quantum version's wave functions, at best, exhibit quasi-periodic motion. This is the result of the kind of evolution isolated, bounded quantum systems undergo. Hence, even if the quasi-periodic wave functions evolve to fill the entire state space, this filling is quite distinct from the kind of state space filling aperiodic trajectories explored in previous chapters.

Moreover, there are interesting effects in the quantum realm that are related to the macroscopic models. For instance, quantum systems exhibit interference effects because isolated wave functions are superpositions of many possible positions and momenta a quantum particle can have. The best-known example is the case of sending an electron through twin slits where the electron passes through both slits and interferes with itself. Sending enough electrons through these twin slits with a detector screen on the other side will build up an unmistakable pattern of interference, as if waves were being sent through the slits.

For the case of classical billiards where the parameters are such that the model exhibits chaos, the corresponding quantized model exhibits almost no quantum interference. On the other hand, if the classical billiard model is exhibiting periodic motion instead, then the well-known interference patterns emerge in the quantized model. Depending on whether the classical billiard is chaotic or not determines whether the quantized quantum analogue exhibits quantum interference.

Relating quantum and classical realms

The kinds of differences seen between the macroscopic and quantized cases raises a question about how the quantum and classical realms are related to each other. I already noted that the kinds of state spaces relevant to quantum mechanics are significantly different than the state spaces of macroscopic physics. Furthermore, the kinds of nonlinear dynamics of the macroscopic world do not exist in the quantum realm. In other words, starting with only the Schrödinger equation and quantum state spaces, it is not possible to produce the equations and state spaces relevant to our macroscopic world.

Then there is the issue of taking a macroscopic model, such as the Chirikov map, and substituting in quantum counterparts for the macroscopic quantities P and θ. What is the connection between the quantum counterparts and the macroscopic variables?

One should say that the connection is "rough," meaning that there is an abrupt shift from the kinds of variables relevant for the quantum realm to the kinds of variables relevant for the macroscopic realm. There is no smooth transition from quantum to classical. Not only this, but it turns out that features called **stability conditions** that do not exist in the quantum realm are required for this shift to happen. This is a form of emergence known as **contextual emergence** (Bishop 2019).

Here is a brief example of the kind of pattern this form of emergence has. The ordered pattern of feather and hair follicles on animal bodies is rather striking, yet the genes giving rise to follicles do not contribute any patterning information. Recall that skin has an epithelial layer forming the epidermis lying on top of the dermis. The epithelial cells in the epidermis have the DNA that produces follicles. During development, the underlying dermis mechanically contracts locally causing the epithelial cells to bend forming little dome shapes creating compressive stress in the overlying epithelial cells. The mechanical contraction from the dermis causes two things to happen with the epithelial cells. First, it breaks the symmetry of the random distribution of epithelial cells producing spatial order. Second the stress forces activate the genetic machinery producing follicles.

In the absence of dermal contraction, the genetic machinery for follicle production never turns on.

The larger-scale mechanical forces provide the stability conditions sufficient to trigger follicle formation in an ordered pattern. True, the underlying genetic machinery provides some of the conditions necessary for follicles—no DNA, no follicles. Nevertheless, it is the larger-scale mechanical forces due to dermal contraction that provide the crucial condition for patterned follicle formation. The genetics of follicle generation is activated and constrained by mechanical forces at a larger spatial scale determining typical distance between neighboring follicles. In this sense, mechanical dermal contraction provides the context for the spatial ordering and production of follicles.

Notice the emergence pattern for follicles does not involve any mysterious brute forces nor presuppose any pre-ordained hierarchy of levels, where smaller-scale factors unidirectionally determine outcomes at larger scales. The pattern reveals how the smaller-scale genetic and the larger-scale mechanical forces work together to produce astounding phenomena such as the striking patterns of bird feathering. Such phenomena do not arise from some set of underlying "governing" or determining laws. The smaller-scale genes have the potential to produce follicles, but this potential is directed by a larger-scale non-genetic condition.

This is the same emergence pattern we see between the quantum and macroscopic realms.

Contextual emergence turns out to be widespread throughout physics but also chemistry, biology, and neuroscience, and has implications for the metaphysics of the sciences and our world (Bishop, Silberstein, and Pexton 2022).

The upshot for quantum chaology is that there is no way for the physics of the quantum realm to produce the macroscopic systems and structures of our world purely from quantum properties and processes alone. On the other hand, there is some evidence that the chaotic macroscopic realm constrains the behaviors of their analog quantum systems. This implies that the chaotic dynamics of macroscopic systems discussed in this book is **contextually emergent**. Here is another important reason why chaotic behavior in the macroscopic realm is not found in the quantum realm. The macroscopic variables and state spaces, nonlinearity, even the determinism necessary for chaos are all contextually emergent in our world.

Interacting quantum systems

How about interacting quantum systems instead of isolated ones? Interacting quantum systems are more complicated and are often called **open quantum systems**. One interesting feature of open systems is that Schrödinger's equation is no longer valid for describing their evolution. Instead, physicists and mathematicians typically turn to so-called **master**

equations to describe evolution. Such equations are used to study the behavior of systems interacting with an environment. An important example in quantum mechanics would be a system, such as a particle, interacting with a measurement device detecting the state of the particle.

In general, the evolution of wave functions in master equations have continuous energy spectra and a continuous energy spectrum is characteristic of macroscopic systems. Moreover, for master equations, two initially close wave functions are not required to remain close to each other, so there is no formal prohibition against wave functions diverging from one another. Interacting quantum systems, then, are better hunting grounds for chaos as observed in the macroscopic realm. Nevertheless, research on interacting quantum systems largely has only uncovered the same kinds of universal statistical characteristics of energy spectra and fluctuations as found in isolated quantum systems.

However, there is one very interesting system with tantalizing behavior. Consider a charged particle in a unit square with periodic boundary conditions, meaning that when the particle reaches the right-hand boundary of the square it reappears at the left-hand boundary and continues its journey (similarly when it reaches the upper boundary it reappears at the lower boundary). Turn an external electromagnetic field on and off quickly to give an occasional kick to the particle. This is like the Chirikov map. Physically, the model represents a charged particle confined to an energy surface shaped like a torus—think donut—that receives kicks from an external magnetic field. For the macroscopic version of this model, the particle trajectories exhibit the stretching and folding process characteristic of chaos and have positive Lyapunov exponents for some parameter values.

For the quantum version of the model, the kick of the electromagnetic field has the effect of taking the quantum labels of wave functions that are initially close together to labels which do not necessarily ever come close again. Labels are parameters in quantum mechanics that define the wave function, such as the angular momentum and spin of a particle. Though somewhat reminiscent of the divergence of classical chaotic trajectories from one another, in the quantum case it is the change in the wave function labels that plays the role of trajectories. This leads to an absolutely continuous **quasi-energy spectrum**, where the quasi-energy is defined as a set of numbers representing the "energy" involved in changing the wave function labels. The particle position becomes unpredictable with respect to the initial wave function labels after long times, and a "distance" between the labels can be defined that increases exponentially with time.

This is the most convincing example that scientists have in quantum chaology of behavior analogous to classical chaos. However, is it the chaos described for macroscopic systems? For instance, the wave function state labels do not change in a continuous "path" the way macroscopic trajectories do. Instead, what happens is there is a jump or discrete transition from one set of labels to another. And, as pointed out earlier, exponential divergence is neither necessary nor sufficient to characterize a system as chaotic. Moreover, quasi-energy is not the same as energy. The name 'quasi-energy' is used in analogy with the energy of particles because it is only due to energy and changes in energy that particles move in macroscopic physics. Quasi-energy is a way of describing wave function label change. True, the behavior of the quantum labels for the kicked particle is irregular, but the actual temporal evolution of the wave functions themselves is regular.

Quantum chaology, while interesting in its own right, is not the same phenomenon we have been exploring in the macroscopic world in this book. Will someone eventually find a convincing example of chaotic dynamics in the quantum realm? Stay tuned!

Chapter 11

Why Chaos Makes a Difference

Surprisingly, chaos turns out to be a scientist's friend. One reason for this is that by applying tools and concepts from the study of chaotic dynamics we can understand and describe unstable systems much better than on any other terms. Moreover, we can forecast them better, too (while some kinds of predictions are severely limited, other kinds of forecasting are highly applicable and useful). Though not all systems in nature are chaotic, the techniques developed for studying chaotic systems and their time series have proved very fruitful in the study of physiology and medicine, biology, ecology, and economics, among other fields.

I will start with some brief summaries of how the study of chaos is making a difference in our knowledge of the world. Then, I will draw out a big picture lesson as I bring this book to a close.

Ensemble forecasting

Your daily weather forecasts, whether on the morning TV or your phone app, report probabilities for rain, snow, and so forth. Where do these probabilities come from? They come from running and rerunning a weather model upwards of fifty times slightly varying the initial data from observations being fed into the model. The probability for rain, say, that the weather forecast reports represents the fraction of those fifty model runs where rain occurs. For instance, if twenty of those runs produce morning showers for Washington D.C., then the meteorologist reports that there is a 40% chance of morning showers for D.C. This is the ensemble used in ensemble forecasting and these techniques are a direct consequence of coming to grips with the sensitive dependence of chaotic dynamics. Contemporary weather forecasts are both much more accurate and more useful for individuals, businesses, and the military than they were before the 1990s.

A further developmental step in ensemble forecasting is to also run multiple models to produce members of the forecasting ensemble. For example, this was the approach applied to the modeling and prediction of hospitalization rates and deaths for Covid-19 to enable local and regional planning and adaptation. However, these forecasts are subject to changes in people's behaviors in a community or changes in government policy. Modeling epidemics requires adjusting for such changes and rerunning the models.

Similar ensemble approaches incorporating multiple models are core to the Intergovernmental Panel on Climate Change modeling assessing future levels of climate change due to global warming. Adding multiple models to the production of our ensemble forecasting system improves the forecasting skill of these systems and enables seasonal and even longer-term predictions of weather and climate systems and their impacts (e.g., Palmer et al. 2004). Drawing on multiple models also implements practical wisdom learned from dealing with faithful model issues (see Chapter 8): Since all models are limited, each model will reflect the values of the model maker to some degree. Hence, the more models forecasters and decision makers use, the better they are able to account for the limitations and value judgments about what factors are more important relative to others represented in any given model.

One advantage of ensemble forecast modeling for the weather is that it can alert meteorologists as to when the weather is in a chaotic state, that is, a state where the stretching and folding dynamics is making the system unstable. Given a set of initial conditions, if our fifty ensemble members do not diverge much from each other, the weather system is quite stable and the uncertainty in our observations will not grow appreciably over the forecasting window (the five-day forecast on the evening news, say). But, if our fifty ensemble members diverge widely from one another over that forecasting window, then we know that the weather is in an unstable state and that the uncertainty in our observations will grow appreciably. That is useful information for meteorologists and affords them strategies for improving the skill of their forecasts during such unstable periods.

Similarly for climate modeling. Even if our climate models cannot capture every detail due to the limitations discussed in previous chapters, they can tell us where in the Earth's atmosphere nonlinearities become important and can amplify small changes (e.g., McIntyre 2021). Hence, even though the amount of water vapor in the atmosphere swamps the amount of greenhouse gases human activity has pumped into the atmosphere, these nonlinearities can amplify the effects of such gases contributing to the global warming disrupting Earth's weather patterns. Furthermore, the heating of water vapor exacerbates hurricanes and other weather phenomena, meaning we should not rest easy thinking that the relative proportions of greenhouse gases to water vapor is nothing to worry about.

Ecological Modeling

Chapter 4 discussed a simple ecological model for species reproduction, the logistic equation, which has rich behavior illustrating chaotic dynamics for appropriate parameter values. Of course, it is difficult to study an isolated population, that is, a population of a species of insect, say, that is not interacting with any other insect species, not to mention plant species, avian predators, and so on. In laboratory environments, it may be possible to create a context where there is just one species and its food source, but it then becomes equally difficult to create conditions that elicit chaotic dynamics in the time series of yearly population growth/decline.

However, it turns out that as we add populations of species to the mathematical model, we get richer chaotic behavior through the interaction among the different species populations. Our mathematical models look more like actual-world ecological communities. Yet, at the same time, the behavior of these models looks as unpredictable as actual communities of organisms, as an early explorer of these models noted (Graves 2022, pp. 70-71). Perhaps it is not surprising that ensemble forecasting has made its way into ecological modeling (e.g., Kranstauber et al. 2022).

Stability of the Solar System

Whether the solar system is stable has been a question since the time of Isaac Newton (1642-1727). The question is tremendously challenging to answer because the Sun and the eight planets all exert gravitational influence on each other. Whereas the problem of two masses interacting by gravity can be easily solved with pencil and paper, for three or more gravitating bodies we generally need computers. Newton famously doubted that the solar system was stable and believed that God had to intervene from time to time to restore the orbits of the planets to order.

Some have tried using the Chirikov map to model the solar system. Imagine that each time a planet comes to the point of closest approach to the Sun or to another planet, it receives a kick due to the force of gravity. We already know that for this map there is the chance for instabilities depending on the strength of kick and the position of the planets. However, to apply the Chirikov map gravity must be active only at these kicks, but we know that gravity can never be turned off or isolated to just one kick at a time. Hence, we cannot settle the status of the solar system's stability this way.

Running models of the solar system using Newton's equation for the force exerted by gravitating bodies, data on the trajectories of the planets, and their masses, all of which are known to very good accuracy, scientists can determine both the past orbits of the planets as well as the future of these orbits. These models predict that roughly eight billion years in the future the planetary orbits will look very similar to now. Given that gravity is a very long range force and over such a long time span small effects have to be included because they potentially can add up to big effects, scientists have had to include all the known comets, asteroids, and planetary moons, as well as passing stars, among others.

So far so good. But now run the same models for slightly varying masses or initial positions of the planets (i.e., vary the initial data within the uncertainties in our measurements). It turns out that over tens of millions of years the planetary orbits diverge tremendously across the various model runs. Our models for the dynamics of the solar system exhibit sensitivity on initial conditions! Analogous to the butterfly effect, this implies that your setting this book down to walk to the kitchen to get a drink can affect the force of gravity enough to shift Jupiter's orbit from one side of the Sun to the other a billion years from now! No planetary collisions are predicted in any of these simulations and the orbits remain ellipses (one can prove that for Newton's gravitational force, orbits must be ellipses). But the positions of the planets on their orbits can vary widely from where they would be if there was no sensitive dependence. Studying chaos has taught us what tell-tale signs to look for.

By varying the initial conditions for our solar system models enough we can see collisions develop in a few of the simulation runs. For example, shift every planet's starting position randomly by a millimeter and on one percent of the runs Mercury ends up colliding with Venus before the death of the Sun. Therefore, we can say the solar system is stable up to the death of the Sun with 99% probability, not unlike our weather forecasts.

Active matter

As is well known, water and oil do not mix—they remain strictly separated in the same container. This separation is an example of **liquid-liquid phase separation**—separation of two or more liquids that is thermodynamically driven. Energy must be added to overcome the separation, as in shaking a bottle of vinaigrette salad dressing to get the separated constituents to mix. As far as molecules are concerned, shaking the bottle represents a lot of energy being dumped into the system breaking down their separation.

Liquid-liquid phase separation is not just relevant for our palates but also crucial to many biological processes. For instance, there are active processes inside cells that can affect the energy level that either maintain or erase such phase separations. To explore this situation, it is possible to set up a biological model using poly(ethylene glycol), known as PEG, with dextran, a polysaccharide derived from glucose, and adding in some protein nanomotors to model active matter (Adkins, et al. 2022). **Active matter**, whether a flock of birds or set of nanomotors, extracts energy from the environment and transforms that energy into useful work (e.g., motion from one location to another).

In the PEG-dextran system, the PEG and dextran can phase separate but, important for modeling cellular processes, their phase boundaries are about a thousand times weaker than oil and water. Hence, it does not take much energy added in to destroy the phase separation and produce a mixture. Enter the little protein nanomotors. When these are added to the PEG-dextran system along with some microtubules, the nanomotors cause the microtubules to start flowing past each other. The astounding thing is that this flow turns out to be chaotic!

The nanomotors are continuously stirring the fluid system. If the PEG and dextran are mixed, the nanomotors cause the system to undergo phase separation so that the PEG forms cohesive droplets within the dextran phase. The nanomotors and microtubules end up in the dextran phase. This means that all the activity is taking place within the dextran separated from the PEG, meaning the PEG droplets are passive subjects to the activity in the dextran. For moderate nanomotor activity, the PEG droplets move around and start coalescing with each other. But if the activity level is too high, the larger PEG droplets break down as fast as the smaller droplets coalesce, leading to an equilibrium where droplets grow no larger.

Meanwhile, the microtubule flows, which are moving the PEG droplets around, in the presence of a wall can actually start climbing up the wall against the force of gravity. It is the chaotic dynamics of the microtubule flows that deform the PEG droplets-dextran interface and the liquid-solid interface between the liquid and the wall. It is these deformations that are crucial to the motions of the droplets and the liquid climbing the wall.

It is the knowledge of chaotic dynamics and understanding what to look for that make it possible for scientists to recognize the crucial contribution of such dynamics to this bio-model. In turn, chaos gives us some insight into how active matter inside cells can move cellular constituents around, up, and down.

Chaos and speech

Human speech is distinctive of all forms of communication among mammals even though it is based on shared acoustic and physiological principles. What makes the difference? Compared to the rest of mammals, nonpathological adult human speech is stable whereas vocalization and phonation in adult species of other mammals is subject to nonlinearities and bifurcations leading to irregular oscillations (Nishimura, et al. 2022). In other words, the difference is that nonpathological human speech typically is not subject to chaotic dynamics!

Most other mammals experience abrupt frequency transitions, irregular oscillations leading to sub-harmonics, and other forms of instability in their vocalization and phonation. This is not simply a matter of having enhanced neural control over our laryngeal muscles, though we do have that compared to the rest of mammals. Physiologically, somewhere along the line humans lost the laryngeal air sacs and the vocal membranes the rest of the great apes have. These physiological features increase the probabilities for nonlinearities in the vocalization and phonation of mammals. And, as you have seen over the course of this book, when nonlinear interactions are present, the possibility for chaos is present. The loss of anatomical complexity in human vocal systems led to much increased stability in our vocalization and phonation. Hence, the conditions that produce nonlinearities leading to sensitive dependence, bifurcations, and aperiodic behavior—chaos—are crucial differentiators for making sounds and communicating in the animal kingdom.

Distinguishing systems and boundaries

As the examples from this chapter show, the study of chaotic dynamics has taught us a lot about how to navigate forecasting in our world and given us the ability to understand features about nature that we otherwise would miss. But wait, there is more to consider. The sensitive dependence characteristic of chaotic systems means that in principle such systems are subject to the minutest of influences, which can be from as far away as is imaginable. But this sensitivity raises a host of questions for how we identify systems and their boundaries, such as how do we make distinctions for systems exhibiting chaotic dynamics?

When scientists model a physical system, they must make distinctions between systems and their environments. However, when the slightest change in the environment of a system can have a significant effect on the system's behavior, the distinction between system and environment breaks down. Maybe when considering a weather system in a chaotic state, it makes sense to identify the flapping of butterfly wings as a disturbance of the environment on the weather system. After all, there are a bunch of butterflies flapping their wings and they seem so tiny. Yet, if they can affect the development of the weather system, it becomes less clear as to how to distinguish the butterflies from the weather system on anything other than pragmatic choice grounds.

Or what about the Earth's surrounding magnetosphere, which exhibits 'space weather' and shields the Earth from lethal solar radiation? Its behavior can impact our chaotic weather system, so is the magnetosphere a distinct system composing part of the weather system's environment or a qualitatively different extension of the weather system? How do scientists make these decisions?

Typically, the identity and individuation of systems and their boundaries revolve around numerical identity and the criteria for individuation and identity through time. In modeling contexts, scientists often draw these distinctions pragmatically because we are always dealing with systems composed of subsystems. There may be some good motivations for drawing a distinction between a subsystem and the rest as its environment, but it is still a somewhat arbitrary distinction. We could have drawn the distinction differently based on our research interests.

All this is to say that it is not so clear how to draw these distinctions when thinking about the interrelatedness of the world in the context of systems behaving chaotically. We are not "carving nature at her joints," as the proverbial saying (or is it a hope?) goes, when it comes to studying chaotic systems. These judgments, encoded in our models, reflect our epistemic access to the systems we are studying and the questions about them scientists want to explore. The sensitive dependence of chaotic systems highlights the epistemic nature and limitations of modeling choices when deciding what is system and what is environment.

Nevertheless, it does not end with mere human judgments about drawing boundaries. Recall that one of the key ingredients for a system to be chaotic is that it be deterministic. For instance, the randomness exhibited by the aperiodic trajectories of the logistic or Chirikov maps is only apparent. One might think that if a mathematician is able to build a chaotic model describing the data in a time series that this implies the system generating the time series data is deterministic. But just as we have problems pulling systems and boundaries apart in the context of chaotic dynamics, we also have problems distinguishing between what is a variable versus a parameter. As you observed earlier, a small change in the parameter for the logistic map can result in the map behaving chaotically or not.

Parameters such as the heat on a system's surface due to its environment may vary over time, producing wide variations in the time evolution of the system variables as well. If this is a nonlinear system, then the distinction between model variables and parameters tends to break down. Scientists usually only consider models and their corresponding systems to be deterministic with respect to the key variables (recall the faithful model assumption in Chapter 8). For the mathematician's model of the time series data, we need to know a lot more than just whether there is sensitive dependence. The mathematician needs to show us that there is the presence of an appropriate stretching and folding mechanism that corresponds to chaotic dynamics. Mere sensitive dependence can be exhibited by nonlinear interactions in a system, but there is no guarantee that such interactions must be deterministic. Hence, making attributions of determinism to the system the mathematician is modeling gets dicey fast.

Chaos warns us we must be careful and cautious with all such judgments distinguishing boundaries from systems, parameters from variables, from what is significant and what can safely be ignored when chaotic dynamics is present.

What about quantum influences?

If the sensitive dependence of chaos can make the boundaries between systems and environments so unclear, is it possible for fluctuations in the quantum realm to affect macroscopic chaotic systems? After all, chaotic dynamics is so sensitive to the smallest influences, it seems reasonable to expect that macroscopic chaotic systems could be influenced by the smallest changes in the quantum realm. If chaos is sensitive to even the smallest effects, it could amplify quantum fluctuations to the macroscopic world, right? The study of chaotic dynamics over the decades has taught us to be alert to such possibilities, but also to be cautious about getting too carried away (e.g., getting too carried away about how much predictability is lost due to chaos).

For instance, one line of argument that sensitive dependence renders chaotic macroscopic systems open to quantum influences comes from thinking about those strange attractors and fractal geometry you met in Chapter 6. The argument goes like this:

1. Chaos in dissipative systems involves strange attractors. As trajectories wind around, they get closer and closer to the heart of these attractors.
2. Strange attractors involve fractal geometry, where the same geometrical structures repeat on smaller scales. These repetitions go on forever.
3. At some point as the length scale decreases enough, the trajectories of the system get so close to the quantum realm that they can now be affected by the smallest quantum fluctuations.

Sounds good until we think about it more closely. After all, strange attractors draw the trajectories of systems towards them such that the state space volume in which trajectories evolve contracts. Nevertheless, fractal geometry only holds for our mathematical models in state space. In the actual world, self-similar structure never exceeds two or three levels of repetition. This means that prefractal geometry never gets close to the length scales for quantum phenomena.

Moreover, the fact that these claims only hold for state space means that our models only include the variables that are relevant for the particular state space in question. There are no quantum variables or effects in the state spaces of our macroscopic models. There is not even any length scale at which quantum effects can be hiding in our mathematical models. Put differently, the state space volume may contract for dissipative models, but there is no quantum realm for the state space trajectories to contract to. Therefore, this line of argument fails to connect macroscopic chaotic dynamics with quantum effects. A lesson from Chapters 8 and 9 is that chaos teaches us to think carefully about our models.

Here is a second line of argument that chaotic macroscopic systems could be influenced by quantum effects:

1. Chaotic dynamics implies that trajectories starting out with slightly different initial conditions will diverge rapidly from one another.
2. Since chaotic dynamics amplifies uncertainties in initial conditions rapidly, any quantum fluctuations affecting the initial conditions will be amplified rapidly.
3. Hence, macroscopic chaotic dynamics could be influenced by quantum effects.

As an example, consider a damped driven pendulum and suppose it is exhibiting chaotic behavior so that it is sensitive to small effects. Further suppose we fire a photon at the pendulum. A photon is a quantum particle, hence in principle if it strikes the pendulum a minute amount of momentum from the photon can be transferred to the pendulum. This sounds just like what this second line of argument is describing.

Nevertheless, the argument runs into problems. The damped driven pendulum is a dissipative system because friction is at work damping its swinging. Mechanisms such as friction will place lower limits on the magnitude quantum effects must have to influence the chaotic dynamics of the pendulum. Here, the devil is in the details, as the saying goes.

Friction is a microscopic process due to the molecular bonds being formed and broken when two surfaces are sliding against each other, as is the case for the pendulum at its pivot point. For our chaotic pendulum to be affected by the quantum effects of the photon, the tiny amount of momentum imparted by the photon must cause at least one molecular bond to form or break that would not have otherwise due simply to the macroscopic mechanical process of the metal surfaces sliding across one another. Hence, the first issue is whether the photon can impart enough momentum to the pendulum to affect at least one molecular bond. Supposing it can, the second issue is whether this small burst of momentum can cause a molecular bond to form or break on a time scale faster than would have been accomplished by the mechanical action of the sliding surfaces.

If you are thinking that the constraints on a small quantum effect such as a photon striking a pendulum sound hard to overcome, you would be right. Though not impossible, quantum effects would have to be quite large and operate quickly enough; otherwise, they will be swamped out by the normal mechanical processes of friction at the pivot point. And recall the earlier discussion of Lyapunov exponents and attractors in Chapter 5. Since trajectory divergence is only an on-average affair, it is possible for a quantum fluctuation to contribute when the trajectories are converging rather than diverging. Under such circumstances the quantum contribution would be lost.

Therefore, the upshot is that it would be very rare for quantum effects to have any impact on macroscopic chaotic systems, if at all. Once again, we see that the study of chaos teaches us to look at the details and carefully assess what is possible.

Chaos is useful

Surprisingly, chaotic dynamics shows up in lots of useful ways in our world. The hullabaloo over unpredictability might lead one to think that chaos is always a problem to be tolerated, but this is not the case. Over the decades there has been much research into how to use chaos in controlled ways.

True, there are some aspects of chaos that are unpredictable or place serious limits on how long scientists can predict system behavior. Nonetheless, it is also the case that the trajectories generated by chaotic signals have well-defined geometrical properties and distribution of frequencies that are stable and repeatable. These present targets for designing systems that take advantage of chaotic dynamics.

In the medical arena, it has been recognized for several decades that many kinds of epileptic episodes occur when normally chaotic processes in brain centers switch to periodic behavior (e.g., Panahi, et al. 2019). There has been much research into understanding the role of chaotic dynamics in the cardiovascular system (e.g., Karavaev, et al. 2019), and how chaos may aid early detection of heart disease (e.g., Singh, et al. 2021). Chaotic dynamics has given us insight into cancer biology, whether it is possible for chaos in protein oscillations, or for chaotic intracellular processes, to play a role in unrestrained cell growth (Uthamacumara and Zenil 2022). This knowledge can provide avenues for medical interventions.

Over the years physicists and engineers have been studying networks of chaotic oscillators for just these purposes. A very useful system to study has been electronic oscillators that can be manipulated simply by the voltage input into the system. When such oscillators are coupled together, they interact in such a way that there is competition between attractors for system behavior and the oscillators repulsing each other. There is a complex dynamics that takes place when chaotic oscillators are trying to find an equilibrium to settle into as a system. Taking advantage of the chaotic network's sensitivity to small changes provides a means for creating a very sensitive sensor for electronic signals. Such systems can also form the basis for distributed computation depending on the kind of voltage input (Minati, et al. 2022).

The sensitive dependence of chaotic systems as well as the intricate structure of their dynamics has inspired study of chaos-based communications systems (e.g., Lau and Tse 2003). For example, manipulating the time series produced by a chaotic system provides a means to code a binary sequence of information (e.g., with positive peaks assigned 1 and negative peaks assigned 0). Since chaos can be produced by simple systems, communications could be performed with very low power apparatus. Additionally, there has been recent work investigating the use of chaotic systems for secure communications (e.g., Zaher and Abu-Rezq 2011).

Recent research indicates that at least 30% of species populations in ecosystems exhibit chaotic dynamics (recall the discussion of the logistic map in Chapter 4). These results have implications for how to consider predictability in the management and conservation of ecosystems and threatened populations (Rogers, Johnson, and Munch 2022). These results require rethinking some conservation strategies.

Finally, there is a handbook for applications of chaos (Skiadas and Skiadas 2016). Chaos is far from a nuisance!

Wisdom of limits

I have given brief examples in this chapter of some of the practical applications of ideas drawn from chaos even if we have no way of proving that we are dealing with the chaos of our mathematical models in these situations. Nonetheless, the examples illustrate how far we can go if we treat these situations **as if** they are chaotic and proceed with caution in case we have judged wrongly.

Yet, the caution we proceed with does not license nonaction when we are making decisions and devising policies based on our data and models. Take the current climate crisis, for instance. We do not know for sure the extent to which our climate models harbor chaotic regimes and we do not know for sure whether Earth's climate system will enter a chaotic regime in the future. None of this gives justification for ignoring the pollution we pour into our atmosphere and environments. We have plenty of evidence that human activity is adversely affecting our planet, particularly the poorest nations and countless endangered species (e.g., Oreskes and Conway 2011; Oreskes 2021; Moellendorf 2022). What we have learned from studying chaos over the past several decades helps us understand the dangers we face and the need for action mitigating all forms of pollution. The examples I have surveyed in this final chapter illustrate that learning about chaos helps us understand more richly the dynamics of the systems we are dealing with—such as environments and the global atmosphere—and what prediction, mitigation, and control can be useful and effective.

Therefore, even if the actual world is not as wild as our mathematical models can be, nevertheless our adventures with chaotic dynamics over the decades have made us better able to understand our world and better prepared to be participants in it. This is an example of learning the wisdom of living within limits, a theme of chaotic dynamics that often goes unnoticed.

Think for a moment how often the theme of limits and their navigation has come up. For instance, Chapter 1 explored some of the predictability limits imposed by chaotic dynamics. This limitation on predictability is partly because there are no perfect measurements. It is not possible to reduce uncertainty to zero, another limitation scientists have learned to work within. And along with these limitations come the finite information storage capacities that scientists also manage as part of their work.

I have noted throughout this book that the rapid growth due to sensitive dependence faces its own limits as space and other resources are always finite (recall the limits to growth reached in the exponential penny map on the checkerboard in Chapter 1). Such rapid growth cannot go on forever. Scientists have learned how to make use of these limitations.

Chaotic dynamics only occurs when the conditions are right for it. Whether in mathematical models or actual-world systems, the conditions leading to chaos are much more restricted than those conditions that do not produce chaos. This is another limitation that scientists have learned how to deal with fruitfully.

In Chapter 3, I noted that nonlinearity is one of the properties all models exhibiting chaos share. Nonlinearity is the ground source for sensitive dependence. In cases where nonlinear effects are important, small changes in the starting state of the model yield rapidly diverging results. Yet, nonlinearities in a system can lead to sensitive dependence without triggering any chaotic behavior. Scientists have learned much about the limitations on when chaotic dynamics emerges and when it does not.

Then there are limitations on scientists' modeling. We cannot model everything, so choices must be made as to which variables and parameters are relevant, what is to be treated as system vs. environment, what questions are relevant to pursue, and so forth.

Chapter 3 also noted that models exhibiting chaos are deterministic. Given the equations defining the model, a deterministic model starting with the same set of initial conditions will produce identical results just like watching your favorite movie repeatedly. Of course, there are legitimate questions about how many systems in the actual world are genuinely deterministic like our mathematical models that all exhibit the unique evolution of a movie. As you saw, this limitation to deterministic models is one reason why chaotic dynamics looks to be limited to macroscopic models and nonexistent for quantum models.

Chaotic dynamics apparently is not limited only to our mathematical models. As Chapter 8 discussed, sensitive dependence and other features of chaos show up in physical systems, weather and climate being some of the most relevant to daily life. The damped driven pendulum is an example of a physical embodiment of chaotic behavior under the right parameter settings. Even though the conditions leading to sensitive dependence only hold for a short time, during this time interval sensitive dependence, limit cycles, and the intricate order of chaotic dynamics are observable in the system. Moreover, as dissipative systems due to friction, such pendulums are observed to harbor strange attractors with their fractal dimensions. Hence, chaos is an actual-world phenomenon, too, though it is much more prevalent in our mathematical models about the world than in the actual world.

Speaking of fractals, another limitation is that such infinitely structured objects are restricted to our mathematical models. The actual world is limited to prefractals at best. Though scientists sometimes will talk in imprecise ways about "fractals" in the physical world, they do know how to deal with the prefractal nature of their subject of study.

Last, but not least, there are the limitations of computers and their ability to represent the actual world. Scientists have learned how to navigate fruitfully within these limitations, too.

Perhaps without realizing it, your journey through this book has been a journey through discovering limits and how scientists have learned to deal with them. Chaos has been a very good teacher for scientists! But there are parallels for all of us here. Just as the scientist must navigate limits on what is knowable about the future of chaotic systems, each person must navigate limits to what can be known and predicted about the future for our lives and loved ones. Although not due to chaotic dynamics, we each navigate human limitations every day. Discussing how scientists work with chaotic systems in this book offers an example of how limitations can be embraced positively and profitably rather than being viewed as obstacles to be overcome. Scientists have learned to work within the limitations associated with chaos fruitfully. That is a lesson all of us can use for how we approach the limitations we face in life.

The work of scientists learning how to work within the limits of chaos is an example for us all. It can inspire and encourage us to learn from and about our limitations and understand more about how we can work within our limits to do good for each other and our world.

Further Reading

Arnold, V. I. (1988), *Geometrical Methods in the Theory of Ordinary Differential Equations*. Springer, New York.

Barbosa, W. A. S., and Gauthier, D. J. (2022), "Learning Spatiotemporal Chaos Using Next-Generation Reservoir Computing," *Chaos: An Interdisciplinary Journal of Nonlinear Science* 32: 093137.

Bishop, R. C. (2005), "Anvil or Onion? Determinism as a Layered Concept,", *Erkenntnis* 63: 55–71.

Bishop, R. C. (2006), "Determinism and Indeterminism," in D. M. Borchert (ed.), Encyclopedia of Philosophy, Second Edition. Farmington Mills, MI: Thomson Gale, Vol. 3, pp. 29-35.

Bishop, R. C. (2019), *The Physics of Emergence*, Institute of Physics Concise Physics Series (San Rafael, CA: Morgan & Claypool).

Bishop, R. C., Silberstein, M., and Pexton, M. (2022) *Emergence in Context: A Science-First Approach to Metaphysics*. Oxford: Oxford University Press.

Crutchfield, J. (1994), "Observing Complexity and the Complexity of Observation," in H. Atmanspacher and G. Dalenoort (eds.), *Inside Versus Outside*. Berlin: Springer-Verlag, pp. 235-72.

Graves Jr., J. L. (2022), *A Voice in the Wilderness: A Pioneering Biologist Explains How Evolution Can Help Us Solve Our Biggest Problems*. Basic Books, New York.

Karavaev, A. S., Ishbulatov, Yu. M., Ponomarenko, V. I., Bezruchko, B. P., Kiselev, A. R., and Prokhorov, M. D. (2019), "Autonomic Control Is a Source of Dynamical Chaos in the Cardiovascular System," *Chaos: An Interdisciplinary Journal of Nonlinear Science* 29: 121101.

Kranstauber, B., Bouten, W., van Gasteren, H., and Shamoun-Baranes, J. (2002), "Ensemble Predictions Are Essential for Accurate Bird Migration Forecasts for Conservation and Flight Safety," *Ecological Solutions and Evidence* 3(3): e12158.

Lau, F. C. M. and Tse, C. K. (2003), *Chaos-Based Digital Communications Systems: Operating Principles, Analysis Methods, and Performance Evaluation*. Berlin: Springer-Verlag.

Lorenz, E. N. (1963), "Deterministic Nonperiodic Flow," *Journal of Atmospheric Science* 20: 131–40.

Minati, L., Li, B., Bartels, J., Li, Z., Frasca, M., and Ito, H. (2022), "Incomplete Synchronization of Chaos Under Frequency-limited Coupling: Observations in Single-transistor Microwave Oscillators," *Chaos, Solitons & Fractals* 165(2): 112854.

Moellendorf, D. (2022), *Mobilizing Hope: Climate Change and Global Poverty*. New York: Oxford University Press.

Nishimura, T., et al. (2022), "Evolutionary Loss of Complexity in Human Vocal Anatomy as an Adaptation for Speech," *Science* 377: 760-763.

Oreskes, N. (2021), *Why Trust Science?* Princeton, N. J.: Princeton University Press.

Oreskes, N., and Conway, E. M. (2011), *Merchants of Doubt: How a Handful of Scientists Obscured the Truth on Issues from Tobacco Smoke to Climate Change*. London: Bloomsbury Publishing.

Panahi, S., Shirzadian, T., Jalili, M., and Jafari, S. (2019), "A New Chaotic Network Model for Epilepsy," *Applied Mathematics and Computation* 346: 395-407.

Palmer, T. N. (2019), "Stochastic Weather and Climate Models," *Nature Reviews Physics* 1: 463-471.

Palmer, T. N., et al. (2004), "Development of a European Multimodel Ensemble System for Seasonal-to-Interannual Prediction (DEMETER)," *Bulletin of the American Meteorological Society* 85: 853-873.

Poincaré, H. (1905), "*Science and Hypothesis: The Value of Science: Science and Method*", tr. George Bruce Halsted (New York and Garrison, NY: The Science Press).

Rogers, T. L., Johnson, B. J., and Munch, S. B. (2022), "Chaos Is Not Rare in Natural Ecosystems," *Nature Ecology & Evolution* 6: 1105-1111.

Ruelle, D., and Takens, F. (1971), "On the Nature of Turbulence," *Communications in Mathematical Physics* 20: 167-92.

Shen, B. L., Wang, M.-H., Yan, P.-C., Yu, H.-P., Song, J., and Da, C. J. (2018), "Stable and Unstable regions of the Lorenz system," *Scientific Reports* 8: 14982.

Singh, R. S., Gelmecha, D. J., Aseffa, D. T., Ayane, T. H., and Sinha, D. K. (2021), "Automated Detection of Normal and Cardiac Heart Disease Using Chaos Attributes and Online Sequential Extreme Learning

Machine," in Manocha, A. K., Jain, S., Singh, M., and Paul, S. (eds.) *Computational Intelligence in Healthcare* (Cham, Switzerland: Springer).

Skiadas, C. H., and Skiadas, C. (2016), *Handbook of Applications of Chaos Theory*. Boca Raton: CRC Press.

Smith, L. A., Ziehmann, C., and Fraedrich, K. (1999), "Uncertainty Dynamics and Predictability in Chaotic Systems," *Quarterly Journal of the Royal Meteorological Society*, 125: 2855–86.

Sokol, J. (2019), "The Hidden Heroines of Chaos," *Quantamagazine*: https://www.quantamagazine.org/the-hidden-heroines-of-chaos-20190520#.

Takens, F. (1981), "Detecting Strange Attractors in Turbulence", in Rand, D.A., and Young, L.-S. (eds.) Dynamical Systems and Turbulence. Lecture Notes in Mathematics, vol. 898. Springer, Berlin.

Thompson, P. D. (1957), "Uncertainty of Initial State as a Factor in the Predictability of Large Scale Atmospheric Flow Patterns," *Tellus* 9: 275–295.

Uthamacumara, A and Zenil, H. (2022), "A Review of Mathematical and Computational Methods in Cancer Dynamics," *Frontiers in Oncology* 12: 850731.

Yang, C., and Wu, C. Q. (2011), "A robust method on estimation of Lyapunov exponents from a noisy time series," *Nonlinear Dynamics* 64: 279-292.

Zaher, A. A., and Abu-Rezq, A. (2011), "On the Design of Chaos-based Secure Communication Systems," *Communications in Nonlinear Science and Numerical Simulation* 16: 3721-3737.

Pay a visit to:

Quick Immersion Series

Visit our WEB:
https://www.quickimmersions.com/

You will get:

+Information of all published books

+News of the books in preparation

+You can subscribe to "A Quick Immersion"

+Links to other spaces of our WEB

+Contact us

+Receive timely information on all our titles